D1032271

Our Evolving Universe is a lucid, non-technical and infectiously enthusiastic introduction to current understanding in astronomy and cosmology. Highly illustrated throughout with the latest colour images from the world's most advanced telescopes, this book also provides a colourful view of our Universe.

Malcolm Longair takes us on a breathtaking tour of the most dramatic recent results astronomers have obtained on the birth of stars, the hunt for black holes and dark matter, on gravitational lensing and the latest tests of the Big Bang. He leads the reader to an understanding of the key questions that future research in astronomy and cosmology must answer. A clear and comprehensive glossary of technical terms is also provided.

For the general reader, student or professional wishing to understand the key questions today's astronomers and cosmologists are trying to answer, this is an invaluable and inspiring read.

Our Evolving Universe

Our Evolving Universe

MALCOLM S. LONGAIR

Jacksonian Professor of Natural Philosophy
Cavendish Laboratory
University of Cambridge

Published by the Press Syndicate of the University of Cambridge
The Pitt Building, Trumpington Street, Cambridge CB2 1RP
40 West 20th Street, New York, NY 10011–4211, USA
10 Stamford Road, Oakleigh, Melbourne 3166, Australia

First published 1996

Printed in Great Britain by Butler & Tanner Ltd, Frome and London

A catalogue record for this book is available from the British Library

Library of Congress cataloguing in publication data

Longair, M. S., 1941–
Our evolving universe/Malcolm S. Longair.
p. cm.
Includes index.
ISBN 0 521 55091 2 (hc)
1. Stars – Evolution. 2. Galaxies – Evolution. 3. Quasars
4. Cosmology. I. Title.
QB806.L66 1996
523.1'1–dc20 95–16415 CIP

ISBN 0 521 55091 2 hardback

For Mark and Sarah

Contents

Preface

'Not another book about the first microsecond of our Universe!', I hear the reader protest. Be reassured – this book is about all the other things which are going on in the Universe around us. In telling that story, however, we are continually driven to ask questions about how all the objects we observe in the Universe today came about and so inevitably we have to look as far into the past as we can in order to disentangle what must have happened.

This turns out to be an incredibly successful and creative quest, far more successful than the pioneers of astrophysics and cosmology of the 1930s and 1940s could possibly have imagined. The development of astronomy and cosmology since the Second World War is one of the most exciting stories of modern science. We have been living through a golden age of astronomy when completely new insights have been gained by totally unexpected means – nobody really expected to discover neutron stars by making observations at long radio wavelengths. Who would have thought that the traces of deuterium present in our environment provide some of the most crucial information about the physics of the Universe when it was only a few minutes old? Who would have guessed fifty years ago that the problems of forming stars would find their solutions in the coldest regions we know of in our Galaxy?

My objective in writing this book has been to write a simple, straightforward, no-nonsense account of the real scientific and intellectual excitement of the whole of modern astronomy. When the words 'evolution' and 'Universe' are mentioned, they are often taken to mean the grand questions of the origins of the Universe itself, but for me they have a different meaning, which I try to bring out in this book. Modern astronomical research spans a vast range of disciplines – the physics of the stars, their birth and death, the physics of the interstellar medium, interstellar chemistry, the formation of galaxies, quasars and supermassive black holes, as well as the questions of the origin of the Universe itself. For me, these different disciplines are part of a single grand design and we need to understand them all to create a convincing picture for the origin and evolution of the different components of our Universe. Even more important, astronomical discoveries are very long range and an innovation in one area can have important and unexpected consequences for other fields.

So the agenda is clear. My aim is to give a coherent account of how all the different disciplines of modern astronomy contribute to the understanding of the origin and evolution of all classes of object in the Universe. For me, each of these areas is just as important as any other. In the process, we have to study the Universe in all the wavebands now available for astronomical observations, from long radio wavelengths to the highest energy γ-rays, each waveband contributing essential information to the story we have to tell. My one regret is that there is not space to describe the enormous contributions of high technology and the genius of experimental and observational innovation. I have included pictures of many of the most important telescopes and I hope these will encourage the enthusiast to follow up the story of these remarkable instruments – without them, we would still be in the dark ages.

The inspiration for this book came from the opportunity I was offered of presenting the 1990 Royal Institution Christmas Lectures for Young People on television. Since that time, the subject has continued to make enormous strides and it has been a challenge to bring this story completely up-to-date, including many of the most recent striking astronomical images. As always, my debt to my wife, Deborah, and our son and daughter, Mark and Sarah, cannot be expressed adequately in a few words. Suffice to say that I had readers like Mark and Sarah in mind when I wrote the book and it is an enormous pleasure to dedicate it to them.

Malcolm Longair
Cambridge,
August 1995

1 The grand design

1.1 The unfolding Universe

One of my favourite stories is the old joke about the unfortunate individual who, somewhat the worse for wear, loses his key on his way home on a very dark night. He is discovered by his friend searching for it underneath a lamp-post. 'Well,' says his friend, 'did you lose your key here, then?' 'Sure, no,' he replies, 'but it's the only place I'd be able to find it, isn't it!' I like this story because it is a precise metaphor for the process of research and discovery in astronomy. Astronomers can only study those regions of the Universe which are revealed to them because they are illuminated by sources of radiation. They simply have to accept what they are offered and make the best use of it in order to understand the nature of astronomical objects. Thus, astronomy differs fundamentally from the laboratory sciences, in which experiments can be performed and repeated under carefully controlled conditions. Astronomers cannot carry out experiments upon the objects they study and the conditions under which they are observed are certainly not carefully controlled. The process of experiment, interpretation and theory is replaced by observation, interpretation and theory. Despite this drawback, astronomical observation has been quite extraordinarily successful in revealing new laws of physics and in leading us to new understandings concerning the origin and evolution of the various components of our Universe. The skill of the astronomer is in devising ways of overcoming the inherent limitations of a purely observational discipline and extracting those universal truths present in the astronomical data.

It always comes as a surprise to me to realise how young astrophysics and cosmology are as scientific disciplines. Modern astronomy is usually considered to have begun with Copernicus' discovery, published in the year of his death, 1543, that the planets move in more or less circular orbits about the Sun (Figure 1.1). The discrepancies between the Copernican circular orbits and the actual orbits of the planets led Tycho Brahe to undertake his magnificent series of observations of the motions of the planets about the Sun. In 1601, Tycho employed as a research assistant the foremost mathematician in Europe, Johannes Kepler, who used Tycho's data to derive what are now called Kepler's laws of planetary motion. These laws led in turn to Newton's discovery in 1665 of the inverse square law of gravity. Thus, from the beginnings of modern science, astronomy has played a central role in the discovery of the fundamental laws of nature.

Progress in understanding the nature of stars and the larger scale structure of the Universe was slow. During the 19th century, however, three important developments took place, which were ultimately to lay the foundations for modern astronomy. One of the big problems was that, up till the 19th century, there was no way of measuring the distance of the stars. The breakthrough came in 1838 when Friedrich Bessel announced the measurement of the trigonometric parallax of the star 61-Cygni. By trigonometric parallax, we mean the apparent movement of the position of the star against the background of very distant stars due to the motion of the Earth about the Sun. The measurement of the parallaxes of stars was, however, a long and arduous business. By 1900, only about 100 parallaxes of nearby stars were known. These observations showed that the nearby stars are objects similar to our own Sun.

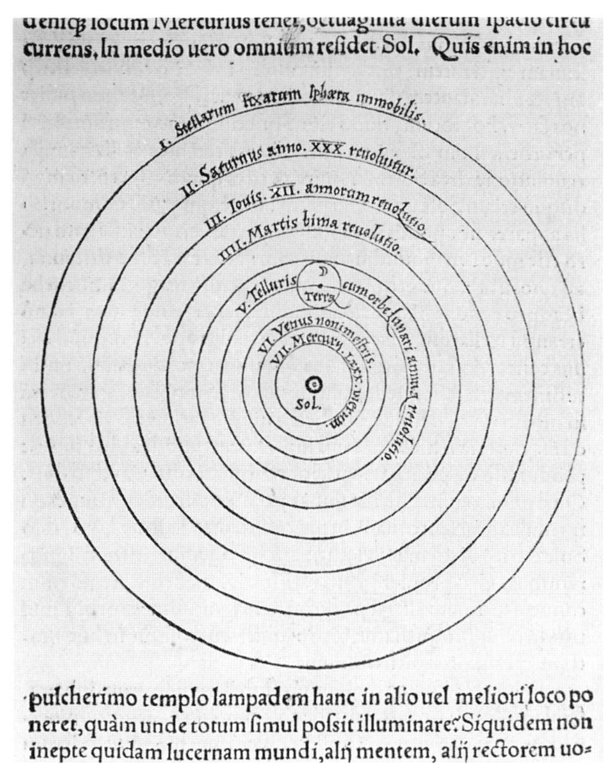
uenię locum Mercurius tenet, octuaginta dierum spacio circu
currens, In medio uero omnium reſidet Sol. Quis enim in hoc

pulcherimo templo lampadem hanc in alio uel meliori loco po
neret, quàm unde totum ſimul poſsit illuminare? Siquidem non
inepte quidam lucernam mundi, alij mentem, alij rectorem uo-

Figure 1.1. The Copernican Universe from Copernicus' treatise *De Revolutionibus Orbium Celestium*, published in Nurnberg in 1543. Copernicus realised that he could account for the motions of the planets, if it is assumed that the Sun is at the centre of the Solar System and the planets move in circular orbits about it. This theory was quite different from the standard picture, attributed to Ptolemy, according to which the Earth is located at the centre of the Universe and the planets move in complex cycloidal orbits about it.

The second development was the discovery of astronomical spectroscopy. In 1814, Josef Fraunhofer made spectroscopic observations of the Sun and discovered a myriad of dark lines in the solar spectrum, which are the signatures of the different chemical elements present in the atmosphere of the Sun (Figure 1.2). In the 1850s, Gustav Kirchhoff was able to identify 30 different elements by the presence of these dark absorption lines in the solar spectrum. The cataloguing of the lines present in the spectra of stars became a major industry in the late 19th century and was to lead to the understanding of the internal structure of the stars during the first three decades of the 20th century.

The third great development was astronomical photography. The discovery of the photographic process was announced almost simultaneously by Daguerre in France and by

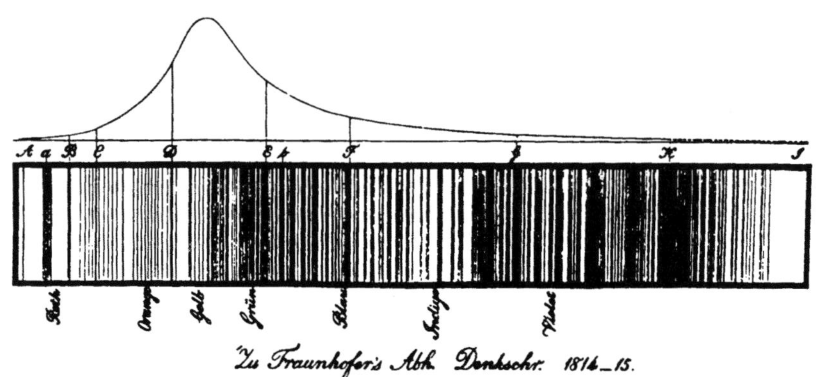

Figure 1.2. Fraunhofer's spectrum of the Sun taken in 1814. His motivation for studying the solar spectrum was his realisation that the most accurate measurements of the refractive indices of glasses should be made with monochromatic light. In his spectroscopic observations of the Sun, he rediscovered the narrow dark lines, already noted by Woolaston in 1802, which provided precisely defined standard wavelengths. He labelled the ten strongest lines in the solar spectrum A, a, B, C, D, E, b, F, G and H and recorded 574 fainter lines between the B and H lines

Fox-Talbot in England in 1839. Among the first pioneers of photography was John Herschel, who took one of the earliest photographs of his father's great 40-foot telescope through a window of his house at Slough in England, just a year before it was dismantled (Figure 1.3). Later in that same year, he took the first photographs onto glass. Although some excellent astronomical photographs were obtained in the middle years of the 19th century, the application of photography to astronomy only became a practical proposition with the development of fast photographic emulsions. In the 1870s, the development of dry gelatin plates reduced the exposure time for terrestrial photography to about 1/15 second. From that time onwards, photography was to become the standard tool of the astronomer. The importance of technological advance in leading to astronomical discoveries will be a recurring theme of our story.

Figure 1.3. One of the earliest photographs of astronomical interest taken by John Herschel in 1839. The photograph shows his father's 12 metre (40-foot) telescope which had a primary mirror of diameter 1.2 m. The exposure was 2 hours.

It was only once the combination of photography and spectroscopy was applied to astronomy that astrophysics, the study of the physics of the stars and gas in cosmic environments, became feasible as a scientific discipline. The great revolutions in physics of the first three decades of the 20th century have their exact counterparts in astronomy, astrophysics and cosmology. By 1939, the basic physics of the stars was understood. The spiral nebulae had been shown to be giant extragalactic systems of stars, similar in structure to our own Galaxy, and Edwin Hubble had shown that the galaxies are all moving apart, what became known as the expansion of the Universe.

Up till the Second World War, astronomy meant optical astronomy. Since 1945, there has been a further revolution in our understanding of the Universe and its contents. By far the most important reason for this has been the opening up of the whole of the electromagnetic spectrum for astronomical observation. First radio astronomy and then X-ray, γ-ray, ultraviolet and infrared astronomy have provided astronomers with completely new views of the

Universe, which are complementary to those obtained with optical telescopes. Each of these new ways of doing astronomy has contributed crucial new information about all classes of object in the Universe – planets, stars, galaxies, clusters of galaxies and the Universe itself. As the new information has been assimilated into the mainstream of astronomy, new insights have been gained into how these different types of object came into being and the role which they play in the evolution of the Universe as a whole. The title of this book *Our Evolving Universe* is designed to encompass how we believe all classes of object in the Universe, including the Universe itself, came about and evolved into the systems we observe today.

None of this new understanding would have been possible without the remarkable technological developments which have made the new astronomy possible. In all wavebands, there have been enormous advances in detector technology and in the techniques of telescope construction. The semiconductor and computer revolutions have been crucial to the advance of both observation and theory. Observatories can now be placed in Earth orbit to study those radiations which are absorbed by the Earth's atmosphere.

At the same time, astronomy has assimilated discoveries made in physics, chemistry and related disciplines into the armoury of weapons used to tackle astrophysical problems. In addition, new types of astrophysics have been developed. For example, it was only with the discovery of interstellar molecules, that the subject of interstellar chemistry was born and, similarly, the discovery of neutron stars as the parent bodies of pulsars led to the intensive study of the properties of matter in bulk at nuclear densities.

Accompanying these great developments, there has been a massive increase in astronomical activity worldwide. To give some measure of this increase, the International Astronomical Union (IAU), which is the internationally recognised organisation for professional astronomers, was founded in 1919. At the first General Assembly, held in Paris in 1922, there were just 200 members from 19 adhering countries. By 1939, the number had risen to about 550 members from 26 countries and the number was roughly the same after the Second World War. By the time of the General Assembly of the IAU in 1991, the number of members had risen to 6700 from 56 adhering countries and the numbers continue to increase. Part of this huge expansion has been associated with the influx of physicists and engineers who, very often, were the pioneers of the new astronomies. In addition, many more large telescopes were made available to astronomers of all persuasions and the population of astronomers has increased to match the growth in observing power. At the same time, the telescopes have become larger, more complex and more expensive, both to build and to operate, with the result that, over the last 30 years, astronomy has become one of the big sciences.

The consequence of these great developments is that we have obtained a completely new understanding of the Universe, in all its aspects. To extend the metaphor introduced at the beginning of this section, we can search for the key using much more powerful telescopes than were available before and these enable us to conduct our search over a much greater volume of the Universe. Even more important, completely new techniques have been discovered for conducting the search. It is still a matter of debate to what extent the astronomers have discovered the key to many of the fundamental problems of modern astrophysics and cosmology, but there is no question that, in conducting the search, they have uncovered a huge range of previously unknown phenomena, and the physical problems to be solved have been much more clearly defined. My objective is to put all these great advances into perspective, concentrating upon what I believe to be the most basic question we have to ask – what are the origins of the diverse objects and structures we observe in the Universe? In the process, we will encounter some of the most profound problems of modern astrophysics and cosmology.

Let us therefore begin with a description of the modern view of the Universe and its contents, as well as indicating how the different astronomies, carried out in the radio, infrared, optical, ultraviolet, X-ray and γ-ray wavebands, contribute to the jigsaw.

1.2 Sizes and distances in the Universe

One of the questions I am asked regularly by non-astronomers is, 'How can astronomers imagine the huge sizes and distances involved in the study of the Universe? How can the human mind comprehend such vast scales?' The answer is that astronomers do not worry about the sizes of the numbers themselves. What is important is the size of one scale in the Universe relative to another. So, let us make a quick guided tour of the whole Universe, starting from our locality and travelling outwards to the most distant objects we can observe in the Universe.

The Solar System Let us start in what we might term our own back garden. What I mean by this term are those regions of the Universe to which we have already sent space vehicles. Much of our Solar System has now been explored by space probes such as the Voyager I and II spacecraft, which returned such fantastic views of the outer planets – Jupiter, Saturn, Uranus, Neptune – and their satellites.

The planets move in more or less circular orbits about the Sun. What is important for our present purpose is the size of the Solar System relative to larger systems in the Universe. First of all, let us compare the distance from the Sun to the Earth with the distance from the Sun to Saturn (Figure 1.4). Saturn is only ten times more distant from the Sun than the

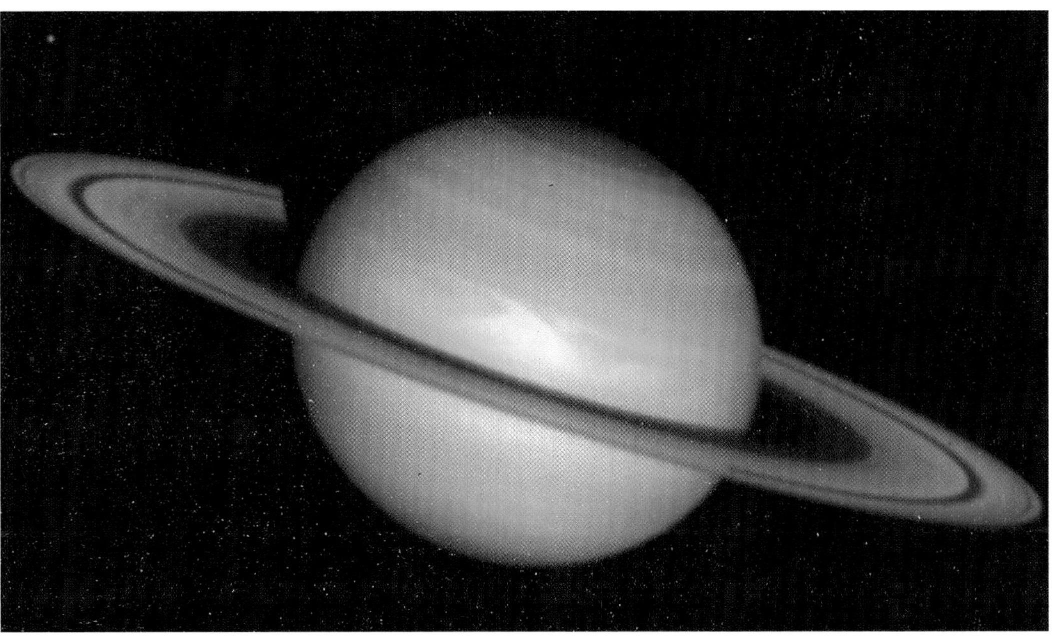

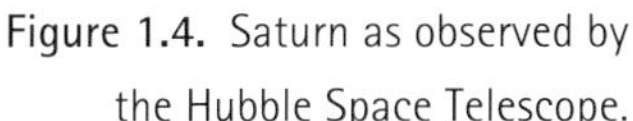

Figure 1.4. Saturn as observed by the Hubble Space Telescope.

Earth. Although the distance in kilometres from the Sun to Saturn is a very large number, roughly 1400 million kilometres, when we think of it as only ten times the distance from the Sun to the Earth, it does not seem such a large distance, relatively speaking. This is the first step in a sequence which will take us out to the edge of the Universe.

The Nearest Stars The next stage in our journey is from our Sun with its nine planets to the nearest stars. If we think in terms of the distance from the Sun to Saturn, rather than in kilometres, the nearest stars are only about 30 000 times more distant – 30 000 is not such a large number. There is another useful way of thinking about the distance of the nearest stars and that is to work out the time it takes light to travel to them from our own Sun. Light travels at a speed of about 300 000 kilometres per second. Therefore, a light signal setting off from the Sun takes only about 8 minutes to travel to the Earth and about 80 minutes to reach Saturn. To reach the nearest star, Alpha Centauri, light has to travel for about about 4.3 years – we say that Alpha Centauri is at a distance of 4.3 light years from the Sun. The light year is a convenient measure of distance in astronomy. To give some idea of how many stars there are nearby, there are about 50 stars within a distance of 17 light years of the Sun. To a rough approximation, there is about one star in every cube of side 7.5 light years in our locality. Thus, stars are very tiny objects indeed, compared with the distances between them.

What do we know about the stars in our neighbourhood? Most of them are cool red stars, similar in many ways to our own Sun, but there are also a few compact hot blue stars known as white dwarfs. Our own Sun is a very average star. The masses of the nearby stars all lie within a quite narrow range, from about two times to about one tenth the mass of our Sun. To find more luminous and massive stars, such as those in the Pleiades (Figure 1.5), we have to search very much larger volumes of the Galaxy, because these are rather rare objects and have very short lifetimes compared with the nearby stars. We will discuss the origin and evolution of stars of all types in Chapter 2.

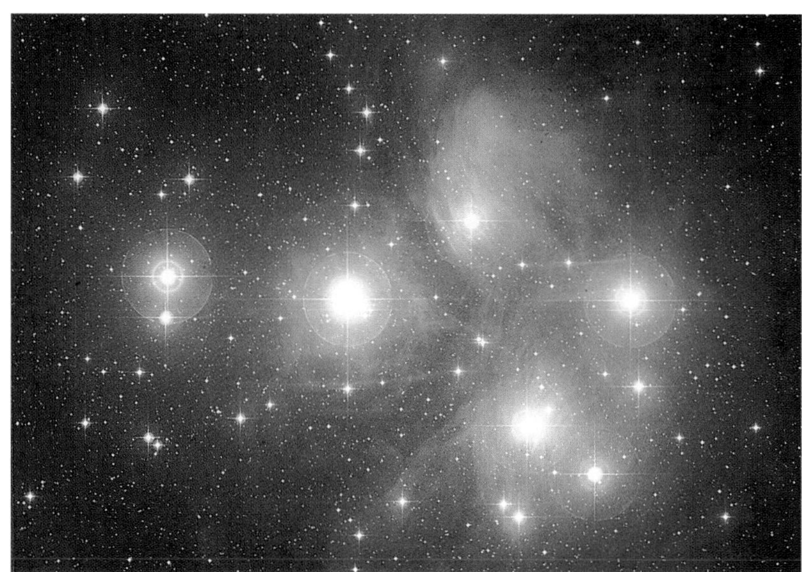

Figure 1.5. The Pleiades star cluster. These blue stars are members of a young star cluster, no more than 20 million years old, and lie at a distance of about 300 light years from the Sun.

The Galaxy Suppose we now go up in scale 20 000 times the distance between the stars – then, we obtain a dramatically different perspective. The stars turn out to be congregated into enormous 'island universes' which are known as *galaxies*. Our own Galaxy, the Milky Way, is probably similar in appearance to the Andromeda Nebula, our nearest giant spiral neighbour in space (Figure 1.6). The stars we observe nearby are only a few of the billions of stars which make up the disc of our Galaxy. The stars of the disc rotate about the centre of the Galaxy. It is the balance between the gravitational attraction of the Galaxy as a whole and the centrifugal force due to their orbital motion about the centre of the Galaxy, which keeps the stars of the disc in more or less circular orbits. In addition to the disc, there is a prominent central bulge of stars. Altogether, our Galaxy must contain about 100 000 000 000 (one hundred billion) stars, a very large number indeed.

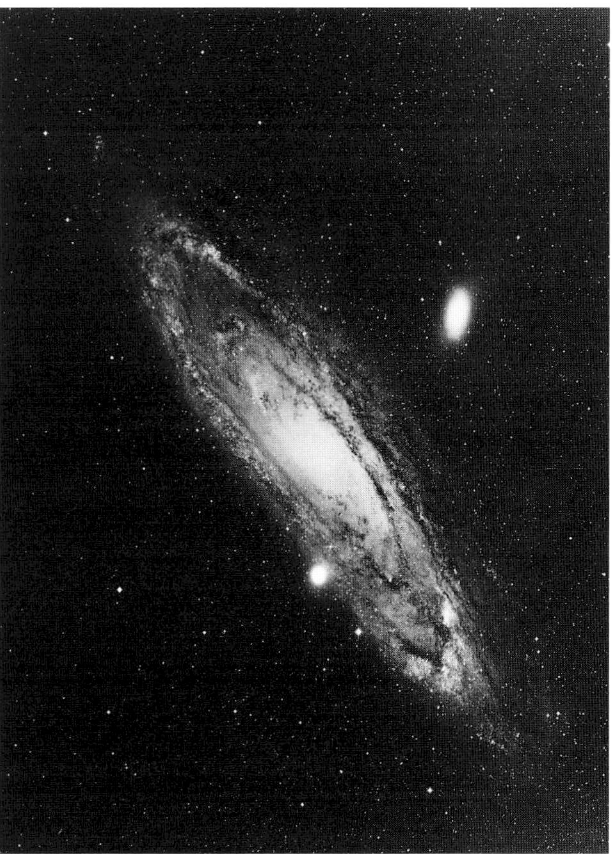

Figure 1.6. The nearest neighbouring giant spiral galaxy to our own Galaxy, the Andromeda Nebula or M31. The Milky Way, which can be seen spanning the whole sky on a clear night (see Figure 1.7), is our own Galaxy, but viewed from the inside. If we were to observe it from a distance, it would probably be similar in appearance to M31.

The problem with observing our own Galaxy is that we are located inside it and so we cannot obtain a bird's eye view of its structure, as we can for distant galaxies. In fact, the Milky Way, which is one of the most spectacular sights of the night sky on a clear dark night, is the disc of our Galaxy, seen edge-on and from the inside. The whole sky is shown in Figure 1.7, in a particular projection which enables us to plot the whole of the celestial sphere on a two-dimensional piece of paper. This projection is known as an Aitoff projection. The plane of the Galaxy, that is, the Milky Way, is placed along the middle of the diagram, with the centre of the Galaxy at the centre of the projection. When we look vertically out of the plane

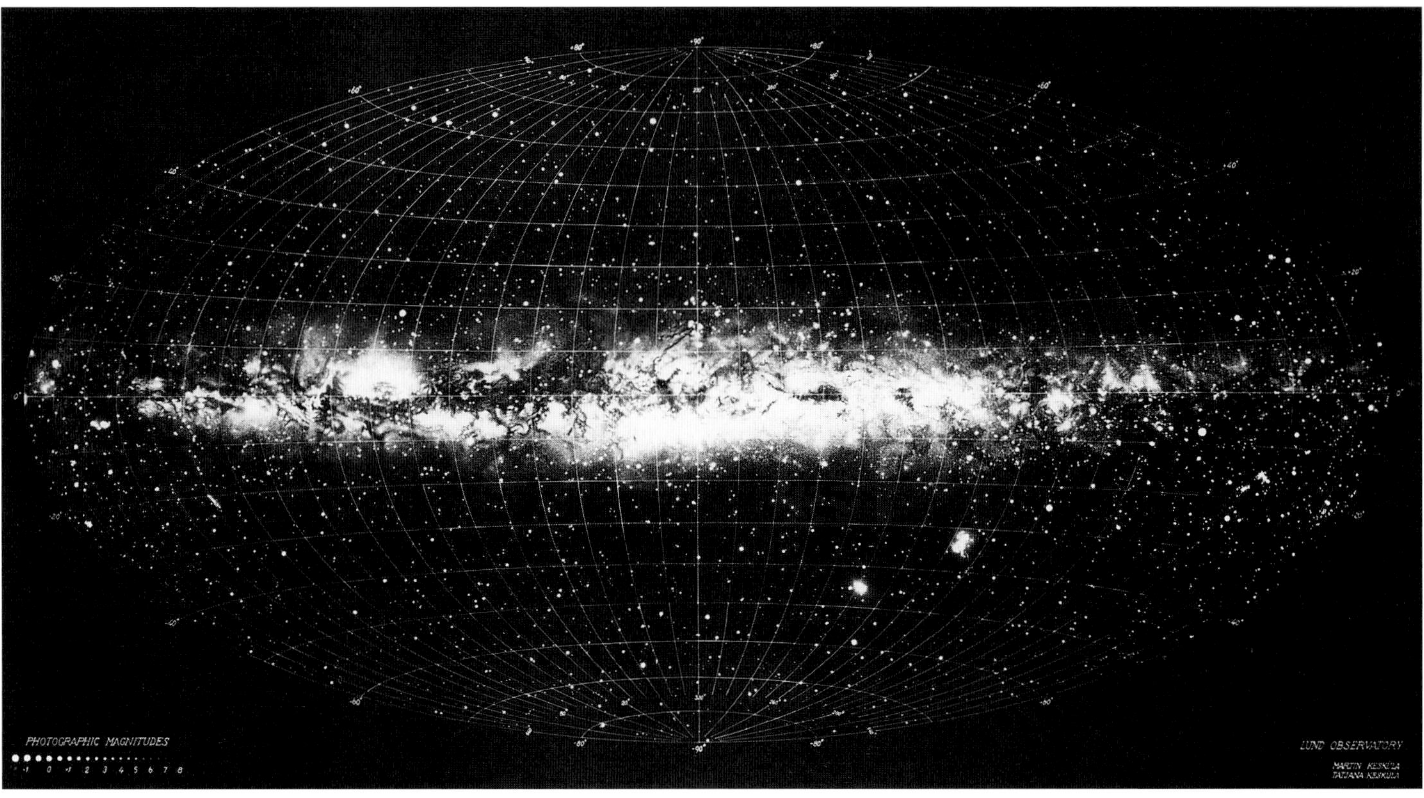

Figure 1.7. A view of the Milky Way showing the whole of the sky projected onto a two-dimensional map. In this picture, the Milky Way runs along the centre of the diagram with the centre of the Galaxy in the centre of the picture. The coordinates have been squashed so that equal areas are preserved. This picture shows that our Galaxy is a flattened disc but the presence of dust prevents us obtaining a clear view of its structure.

of the Galaxy in the northern hemisphere, we look in the direction of the *North Galactic Pole* and it is placed at the top of the diagram. The corresponding *South Galactic Pole* is located at the bottom of the diagram. Areas are preserved in this projection but, to achieve this, the coordinates are rather badly bent towards the poles and this is why the picture of the sky shown in Figure 1.7 has a rather squashed appearance. This is a very useful way of representing the whole sky on a single flat sheet of paper and we will look at many maps of the sky made in different wavebands in the same projection.

Figure 1.7 does not look particularly like the Andromeda Nebula and the reason for this is apparent. It can be seen that there are dark patches on the picture and these are opaque dust clouds which prevent us observing optically to large distances in these directions. In fact, not only are there dense dust clouds, there is also diffuse interstellar dust distributed in the space between the stars and this prevents us observing optically the structure of our Galaxy. This posed a great problem for the pioneers of Galactic structure because it was only in the 1930s that the obscuring effect of diffuse interstellar dust was fully appreciated. There is, however, a way of obtaining an unobscured picture of the Galaxy nowadays. As we will explain in Section 2.4, dust becomes transparent at infrared wavelengths of about 1 to 3 μm (microns), wavelengths only about four times longer than typical optical wavelengths. A beautiful map of the whole sky has been made by the Cosmic Background Explorer (COBE) satellite at these wavelengths and it is shown in the same Aitoff projection in Figure 1.8. This map of the Galaxy shows clearly that it consists of a thin, flattened disc, and has a prominent central bulge, similar to that observed in spiral galaxies such as the Andromeda Nebula.

Figure 1.8. An image of the Galaxy as observed by the Cosmic Background Explorer (COBE) in the infrared wavebands, 1.2 to 3.4 μm. The thin disc of our Galaxy and the central bulge can be observed, because dust becomes transparent in the infrared region of the spectrum.

The galaxies In 1926, Edwin Hubble showed conclusively that spiral nebulae, such as the Andromeda Nebula, are 'island universes', similar to our own Galaxy. Typically, galaxies have masses similar to our own Galaxy and the Andromeda Nebula but there is a wide range of masses about this mean value, from about 10 million to 10 thousand billion times the mass of the Sun in the most extreme cases. The galaxies are the building blocks of the Universe and they define its large scale structure. The vast majority of galaxies are called *normal galaxies* and they come in two broad categories. One of these consists of the *spiral galaxies* and our own Galaxy and M31 belong to this class. The characteristic feature of the spiral galaxies is the presence of spiral arms, which are defined by young hot blue stars, gas clouds and dust. In many spiral galaxies, the spiral arms originate from within the central bulge and these are called *ordinary spiral galaxies*. Equally numerous are the *barred spiral galaxies*, in which the central bulge is replaced by a 'bar' of stars (Figure 1.9).

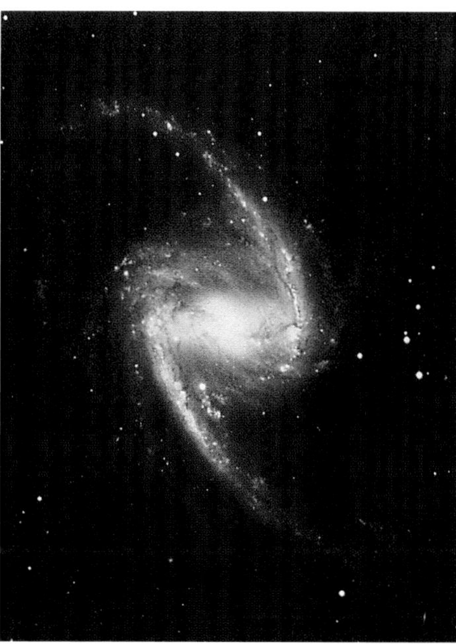

Figure 1.9. The barred spiral galaxy NGC 1365.

The second major class of galaxies consists of the elliptical galaxies, which are spheroidal systems. They do not contain discs of stars and the distribution of light is smooth (Figure 1.10). The properties of the elliptical galaxies are similar to those of the bulges of spiral galaxies. There are also galaxies intermediate in appearance between the spiral and elliptical classes and these are known as the *S0* or lenticular galaxies. They appear to have a disc and bulge structure, but there is no evidence for any spiral structure – it is as if a spiral galaxy had been stripped of all its gas and dust and, indeed, this is one of the more promising theories of their origin.

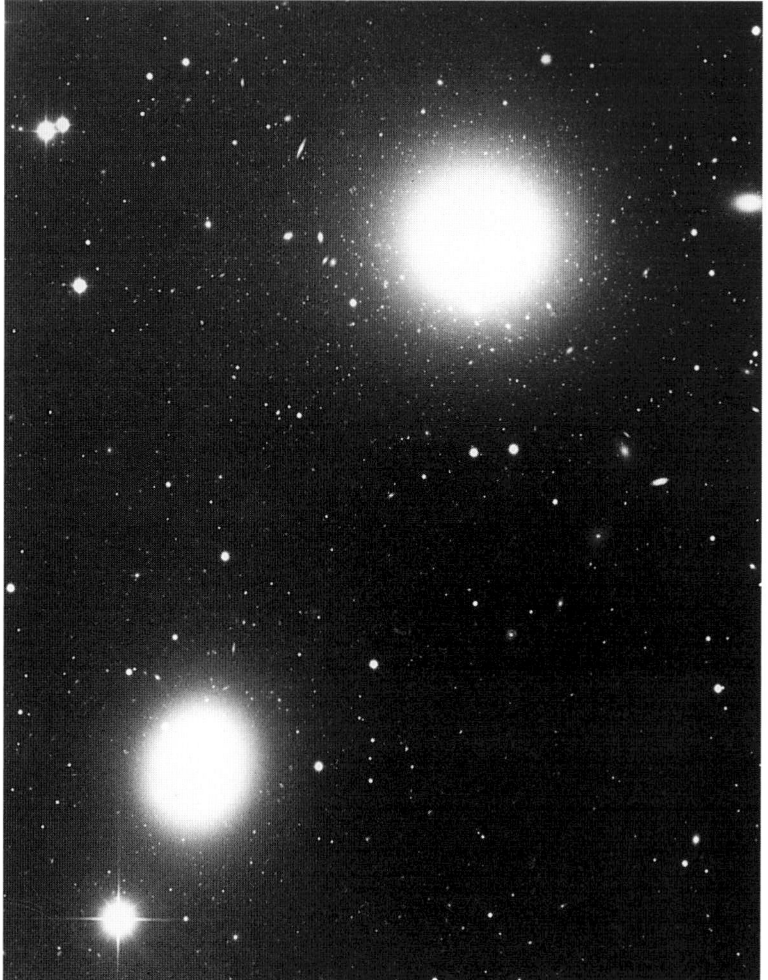

Figure 1.10. The elliptical galaxies NGC 1399 and 1404, which are members of the Pavo cluster of galaxies.

Finally, a few galaxies belong to the small class of *irregular galaxies*. Many of these are low mass galaxies and have an irregular appearance, occasionally with some vestiges of spiral structure. Among this class, there are peculiar galaxies such as the Cartwheel, shown in Figure 1.11, which seems to consist of a ring of stars. It is probable that the galaxy has been involved in a collision with one of its neighbours and has left behind a ring of young stars. Collisions between galaxies are probably the most common cause of such peculiar galaxies.

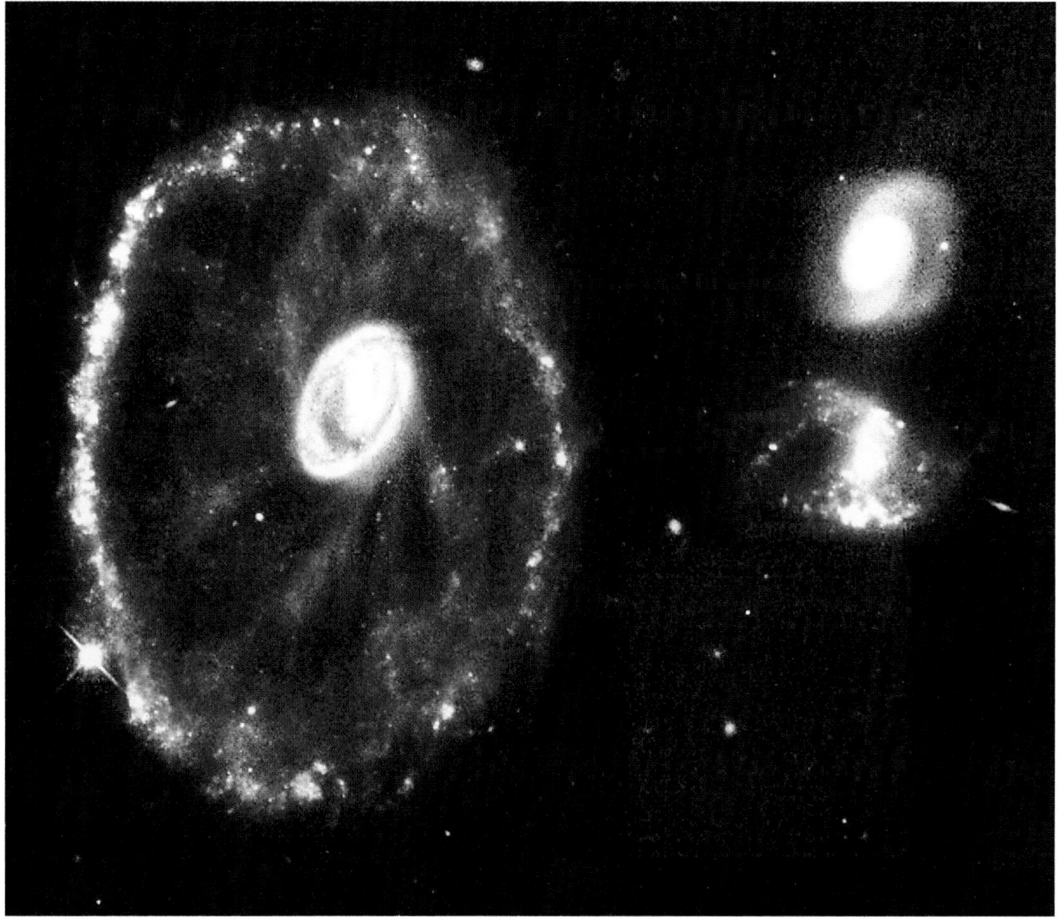

Figure 1.11. The peculiar galaxy known as the Cartwheel as observed by the Hubble Space Telescope. Its strange appearance is almost certainly due to a recent collision or to a strong interaction with one of its nearby companions.

Active galaxies and quasars All the galaxies in the Universe can be classified into the general categories described above and we will study their origin and evolution in Chapter 4. There are, however, galaxies in which much more dramatic events are observed to take place. A good example is the galaxy NGC 4151. At first sight, this galaxy appears to be a rather harmless spiral galaxy (Figure 1.12). When a short exposure photograph of the galaxy is taken, however, something very dramatic seems to be taking place in its nucleus (Figure 1.13). There seems to be a 'star' located precisely in the centre of the galaxy. It cannot, however, be any normal type of star at all, because it is at the same distance as the galaxy and so must be very much more luminous than the most luminous stars we know of. What makes matters even more intriguing is the fact that these star-like objects are known to be variable in intensity which, as we will show in Chapter 3, means that they must be very compact indeed. This is an example of the class of galaxies which possess what are known as *active galactic nuclei*.

In the case of NGC 4151, the active nucleus is much fainter than the galaxy but, in very rare cases, the opposite is found – the nucleus far outshines the total light of the galaxy and these are the objects known as *quasi-stellar objects* or *quasars*. Figure 1.14 is a photograph of the brightest quasar in the sky, 3C 273. The faint smudges towards the bottom of the picture are companion galaxies at the same distance as the quasar. Evidently, the nucleus outshines

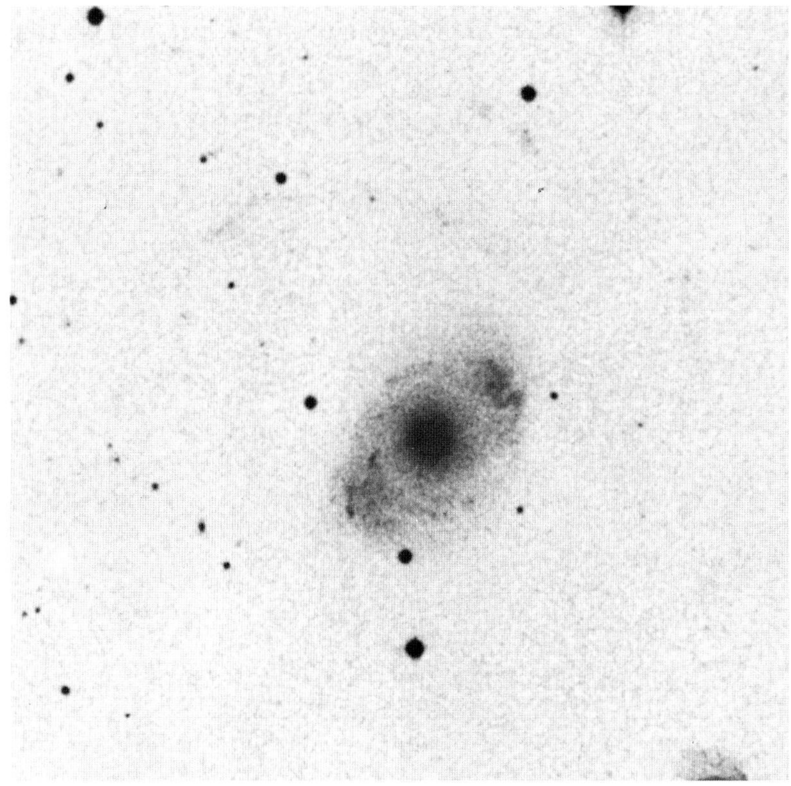

Figure 1.12. A long exposure photograph of the galaxy NGC 4151. The galaxy appears to be a spiral galaxy.

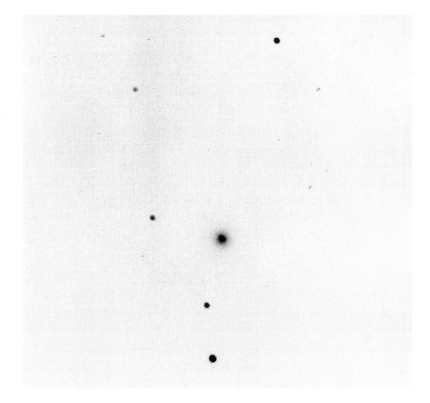

Figure 1.13. A short exposure photograph of NGC 4151. The nucleus of the galaxy appears to be stellar. The fact that the nucleus is very compact is confirmed by the observation that its brightness varies.

the underlying galaxy by a factor of about 1000 in luminosity. The quasars and their violent close relatives, the objects known as *BL-Lac objects* and *blazars*, are the most powerful sources of energy that we know of anywhere in the Universe and we will tackle the problems they pose in Chapter 3. For the moment, we note that most galaxies are more or less normal galaxies and that only very rarely do we come across exotic objects like the quasars, which have quite different properties.

Clusters of galaxies Let us continue our journey going to larger and larger scales in the Universe. We have already stated that the galaxies are the building blocks of the Universe and so, to make a map of the structure of the Universe on the very largest scale, we need a three-dimensional map of where the galaxies are located. It turns out that the galaxies are far from randomly distributed in space. They generally form rather irregular structures, as we will see in a moment, but there exist some large systems of galaxies, the largest bound systems that we know of in the Universe, which are called clusters of galaxies. In the great clusters of galaxies, there may be many thousands of galaxies and the cluster is held together

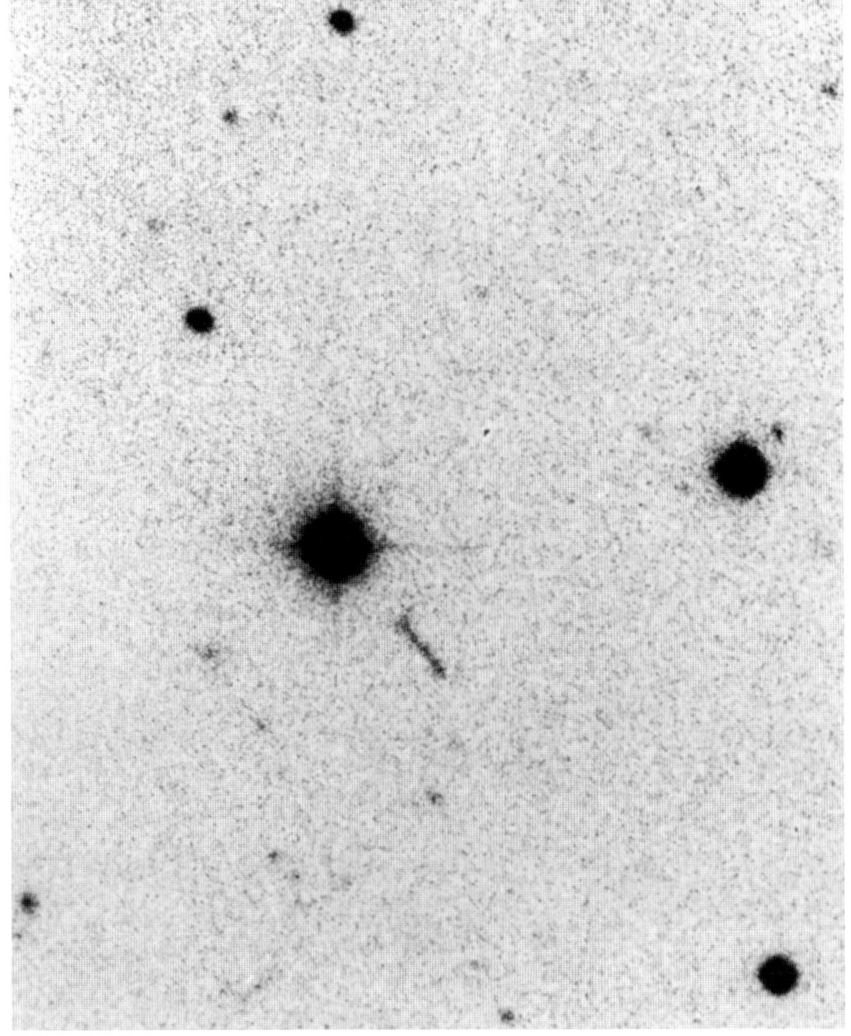

Figure 1.14. The quasar 3C 273 is the brightest object of this class known. Although it looks like a star, it is in fact the active nucleus of a very distant galaxy. The faint smudges seen to the south of the quasar are galaxies at the same distance as the quasar. The nucleus outshines the galaxy by a factor of about 1000 in luminosity. 3C 273 possesses a remarkable optical jet which can be seen to the bottom right of the image of the quasar.

by the mutual gravitational attraction of the galaxies for one another. Figure 1.15 is a picture of the Pavo cluster of galaxies and our own Galaxy would not be a particularly conspicuous member, certainly not nearly as prominent as the monster, giant elliptical galaxy in the centre of the cluster. The typical size of one of the great regular clusters, such as the Pavo cluster, is about 50 times the size of our Galaxy.

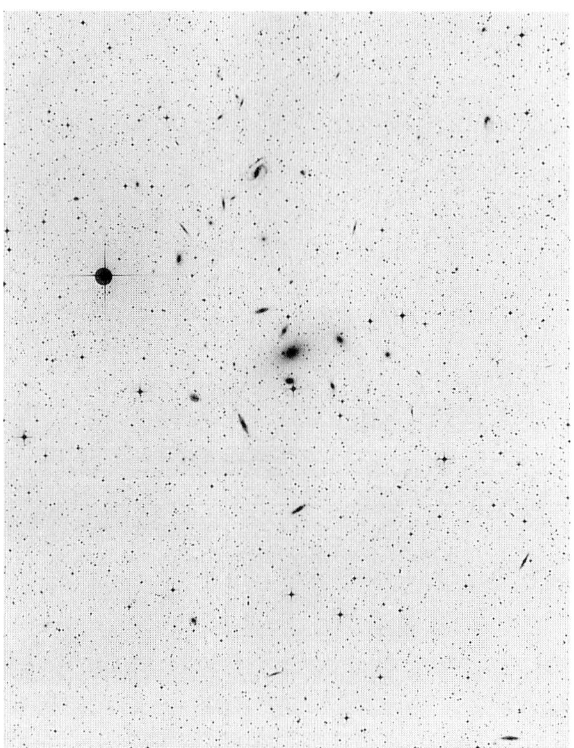

Figure 1.15. The Pavo cluster of galaxies. In this image, our own Galaxy would be similar to the disc galaxy seen towards the bottom of the photograph.

The large scale structure of the Universe The great clusters are spectacular objects, but what about the majority of galaxies which do not lie in great clusters – how are they distributed in the Universe? This is one of the most exciting and important areas of modern cosmology. The distribution of galaxies on scales larger than clusters of galaxies does not seem to be particularly smooth. Figure 1.16, which was created by James Peebles and his colleagues at Princeton University, USA, from galaxy counts made at the Lick Observatory in California, is a picture of the whole of the Northern Galactic Hemisphere, that is, the whole region of northern sky away from the Milky Way, with all the stars belonging to our own Galaxy removed. To create this picture, over one million galaxies were counted to the limits of the plates of the Lick astrographic survey which corresponded to an apparent magnitude of 18.5. This is a view of the Universe on the very grandest scale. Individual clusters of galaxies are the bright areas seen in Figure 1.16; for example, the prominent bright spot close to the centre of the picture is the Coma cluster of galaxies. The galaxies are clearly not uniformly distributed on scales greater than those of clusters of galaxies – there seem to be holes, filaments and sheets of galaxies.

The problem with Figure 1.16 is that it is a two-dimensional projection of the three-dimensional distribution of galaxies. The true nature of their distribution is most beautifully illustrated by the results of the Harvard–Smithsonian Center for Astrophysics Survey of

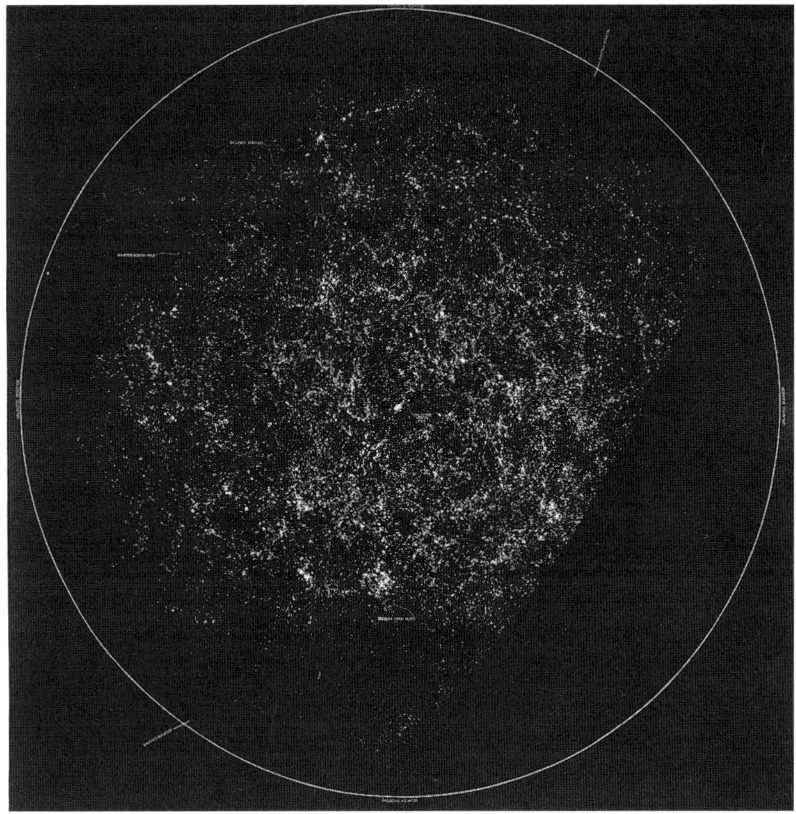

Figure 1.16. A picture of the distribution of galaxies in the Northern Galactic Hemisphere. The centre of the diagram corresponds to looking vertically out of the plane of our Galaxy, and the outer white circle corresponds to looking through the Galactic plane. The coordinate system has been chosen so that equal areas are preserved. The lack of galaxies towards the edge of the diagram is due to obscuring dust in the plane of our Galaxy. The 'bite' out of the diagram at the bottom right occurs because that area of sky was not surveyed.

Galaxies. In this survey, distances as well as positions on the sky will eventually be measured for over 30 000 galaxies. Figure 1.17 is one of the most important results of that survey and shows the distribution of galaxies within a 'slice' through the nearby Universe. Our own Galaxy lies at the centre of the diagram. Within this slice, the Harvard astronomers, Margaret Geller and John Huchra, have plotted the radial velocities of the galaxies away from our own Galaxy as measures of their distances. In Chapter 4, we will show that the radial velocities of galaxies are proportional to their distances from our Galaxy. This was Hubble's great discovery of 1929 and is known as Hubble's law.

If the galaxies were uniformly distributed in the Universe, the points would be uniformly distributed within the bounding circle of Figure 1.17. It is obvious that the galaxies are far from uniformly distributed in space. Let us consider some of the more prominent features of their distribution. There are several 'lines of galaxies' pointing towards our own Galaxy at the centre of the diagram. These are the signatures of rich clusters of galaxies, similar to the Pavo cluster of galaxies shown in Figure 1.15. These are gravitationally bound systems and the large dispersion of velocities of the galaxies within the bound cluster is responsible for stretching out the estimated distances of the galaxies in the radial direction, despite the fact that they are all actually at the same physical distance. These features are sometimes referred to as 'the fingers of God'. Other large scale features include sheets and filaments of galaxies and huge holes which are often called *voids*. There is a huge chain of galaxies in the upper part of the diagram which stretches across almost the entire region surveyed – this is sometimes referred to as the 'Great Wall'. Figure 1.17 illustrates the gross irregularities present

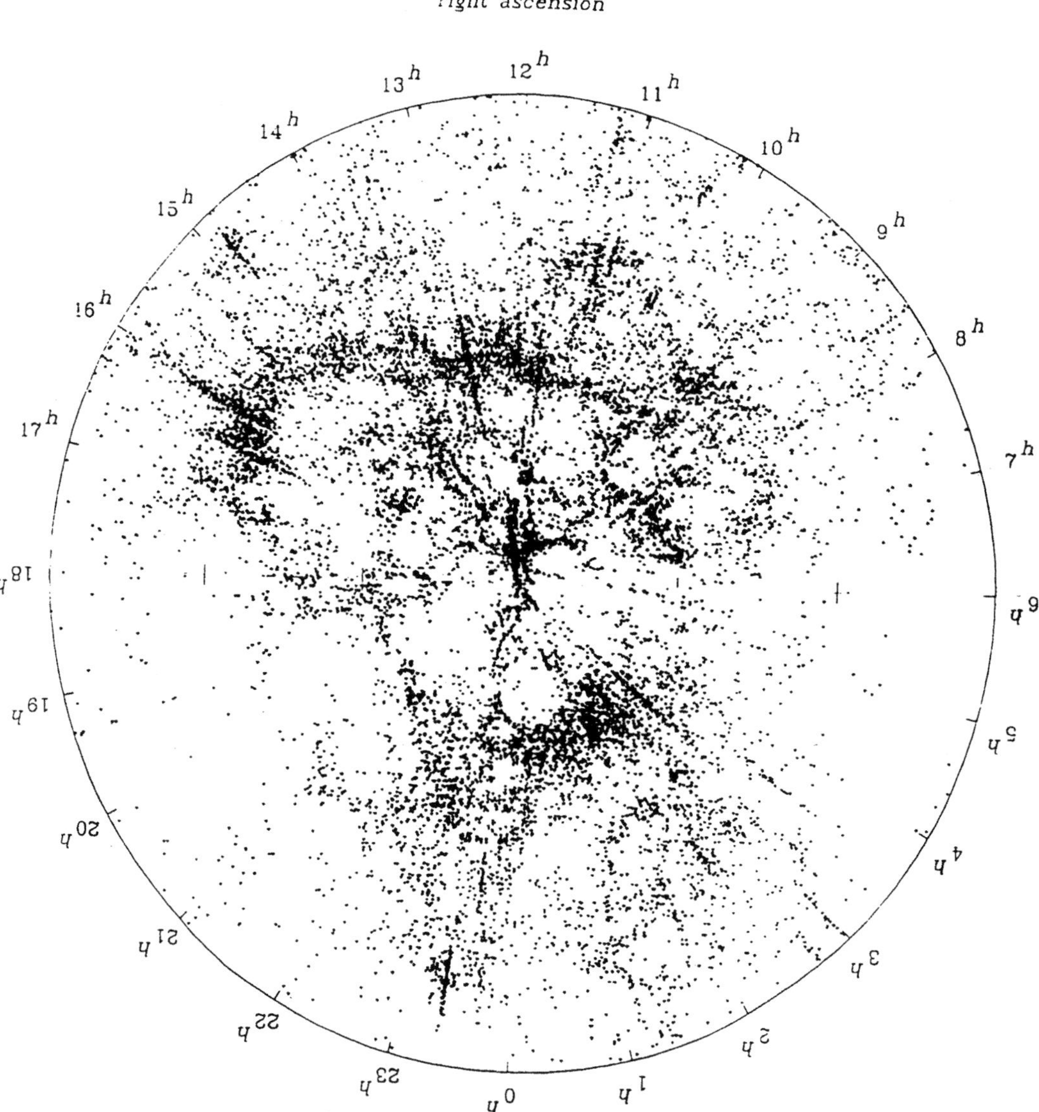

Figure 1.17. The distribution of galaxies in the nearby Universe as derived from the Harvard–Smithsonian Center for Astrophysics Survey of Galaxies. The map contains over 14 000 galaxies which form a complete statistical sample between declinations 8.5° and 44.5°. All the galaxies have velocities of recession less than 15 000 km s^{-1}. The galaxies within this slice around the sky have been projected onto a plane to show the large scale features of their distribution. Rich clusters of galaxies, which have large internal velocity dispersions, appear as 'fingers' pointing radially towards our Galaxy, which is at the centre of the diagram. The distribution is grossly non-uniform with huge sheets, filaments and voids in the distribution of galaxies. The radius of the outer bounding circle is about 500 million light years.

in the distribution of galaxies and it is one of the greatest challenges of modern astrophysical cosmology to account for the origin of these structu.re

What is the best way of envisaging the distribution of galaxies in space? Richard Gott and his colleagues at Princeton University have provided an elegant and simple answer. We can think of the Universe as being rather like a sponge. The material of a sponge is all joined together or else it would fall apart. What is more remarkable is the fact that the holes in the sponge are all joined together too! This type of topology is possible in three dimensions but not in two. This is how the matter seems to be distributed in the Universe. The holes are all connected and the distribution of galaxies, corresponding to the material of the sponge, is all connected as well. The rich clusters of galaxies tend to lie at the vertices of the sponge, regions where various bits of the sponge come together. It is one of the challenges of theory to explain how this topology of the distribution of galaxies came about.

We return to the question of the relative sizes of structures in the Universe. The next scale-size up from that of a cluster of galaxies is the size of the large voids seen in the distribution of galaxies. The largest voids that have been detected so far are about 50 times the size of a cluster of galaxies and this is also the typical distance between rich clusters of galaxies.

The size of the Universe Finally, what about the size of Universe itself? The size of the Universe is only about 50 times the size of one of the large voids. It may come as a surprise that the Universe is so small! The point is that we have to be careful about what we mean by the size of the Universe. One of the fundamental laws of physics is that the speed of light always has the same very large value of about 300 000 km s^{-1} when it travels through a vacuum. Therefore, as we look further and further away, it takes longer and longer for the light to reach us. As a result, when we observe very distant objects, we do not study the Universe as it is now, but as it was when the light was emitted, which was in the distant past.

Let us be more precise about what I mean by the 'size of the Universe'. We will show in Chapters 4 and 5 that we live in a Universe which has expanded from a very compact state indeed, and that all the galaxies are moving apart as a result of that initial explosion – this is the standard 'Big Bang' picture of the Universe. Now, if we study the Universe, not right back to its beginning, but only to the time when it was half its present age, we are no longer looking at the Universe as it is now because it was much younger then and will certainly be significantly different from the Universe we observe now – the clusters of galaxies, the galaxies and the stars in them will all be at most half their present ages. What we mean by the size of the Universe now is the distance out to which we can see the Universe more or less as it is today and I have adopted as a measure of this the distance light has travelled since the Universe was about half its present age – that distance is roughly 50 times the size of the large voids.

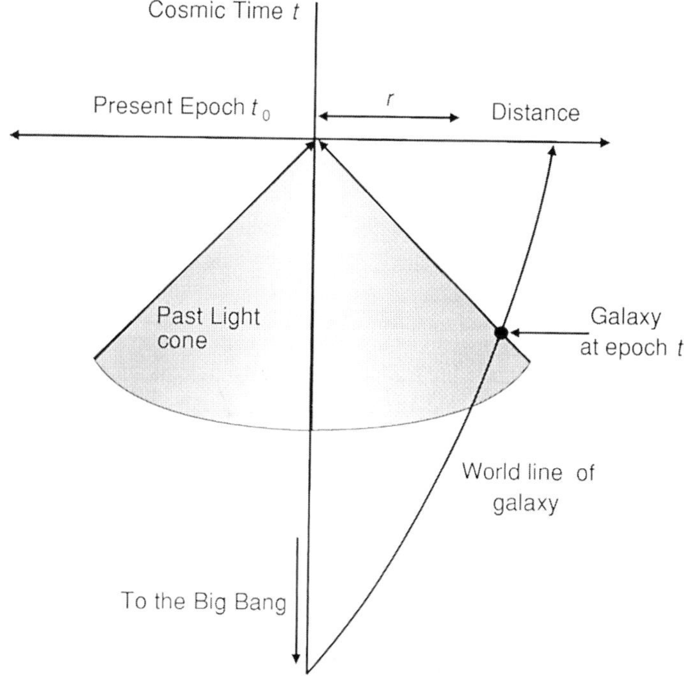

Figure 1.18. A simple space – time diagram illustrating what is meant by the 'size of the Universe'. Time runs up the diagram and distance across it. Our Galaxy travels up the vertical axis at $r = 0$ from the Big Bang at $t = 0$ to the present age of the Universe, t_0. When we observe distant objects, we look into the past. Eventually, when we look very far away, we see the Universe as it was when it was much younger than it is now.

This idea is illustrated in Figure 1.18 which is a simple space–time diagram for the Universe. Time is plotted up the vertical axis with the origin at the Big Bang. Along the horizontal axis, distance from our Galaxy is plotted. In this picture, the path of our Galaxy through space–time is represented by the vertical axis, corresponding to $r = 0$. The paths of galaxies on the diagram all come together in the Big Bang and diverge up the diagram – this is simply the expansion of the distribution of galaxies and the trajectory of one of them is shown. Light travels to the Earth at the speed of light and this is shown as a cone, converging on the Earth, at the present epoch, from the past. All information about the Universe comes to us along this cone, which is known as our *past light cone*. We observe the galaxy shown at the time corresponding to the point at which the trajectory of the galaxy intersects our past light cone. What I mean by the size of the Universe can be thought of as the distance light travels in half the age of the Universe, as illustrated in the diagram.

It can be seen from Figure 1.18 that we can only observe regions of the Universe which are within our past light cone and consequently there are regions which we cannot possibly observe at the present time – light cannot reach us from these regions in the time since the light was emitted. If we look further and further away, we indeed observe more and more distant regions of the Universe, but not as they are now, but as they were in the more and more distant past. This is why observational cosmology is feasible at all. Assuming distant parts of the Universe had a similar history to the regions in our locality, we can, in principle, determine the history of the Universe directly by observation. How far back we can trace this history depends upon the objects we choose to study. The most luminous galaxies and quasars, which can be observed with the present generation of large telescopes, emitted their light when the Universe was less than one-fifth its present age. Unfortunately, it is only the most luminous objects which can be observed back to those early epochs. With telescopes such as the Hubble Space Telescope and the Keck 10-metre telescope located on Mauna Kea in Hawaii, it is just possible to study ordinary galaxies back to the time when the Universe was about half its present age.

We have now travelled from our back yard, our own Solar System, right to the edge of the Universe in only six modest steps and we have not been talking about huge numbers at all. Rather, we have needed only relatively small numbers, which were never any greater than 30 000. These sizes and dimensions are summarised in Table 1.1. It is an interesting scientific point that, when scientists are faced with the problem of having to deal with a very wide range of scales, they do not think linearly in terms of all the zeros on the ends of the numbers, but rather they think logarithmically, that is, of one scale relative to another. I illustrate this idea schematically in Figure 1.19, in which the scale of distance along the horizontal axis is x, the logarithm of the distance. It can be seen that, by taking logarithms, the whole Universe can be squashed onto a single sheet of paper.

1.3 The nature of light and the electromagnetic spectrum

The description we have presented of the contents and large scale structure of the Universe has been based upon optical observations. This introduces a bias into the picture because light is only one particular type of wave, which can be used nowadays to observe the Universe. Light is a form of what is known as electromagnetic radiation, meaning the waves associated with oscillating electric and magnetic fields. The discovery that light is a form of electromagnetic

Table 1.1. *Sizes and distances in the Universe*

Scale or distance	Relative size	Distance in kilometres	Distance in light years
Sun to Earth	A	1.5×10^{8}	0.000 015
Sun to Saturn	B = 10 × A	1.5×10^{9}	0.000 15
The nearest stars	C = 30 000 × B	4.5×10^{13}	4.7
Our Galaxy	D = 20 000 × C	1.0×10^{18}	100 000
Clusters of galaxies	E = 50 × D	5.0×10^{19}	5 000 000†
Giant voids	F = 50 × E	2.5×10^{21}	250 000 000†
The Universe as it is today	G = 50 × F	1.3×10^{23}	14 000 000 000†

† These distances are uncertain by about ±50%.

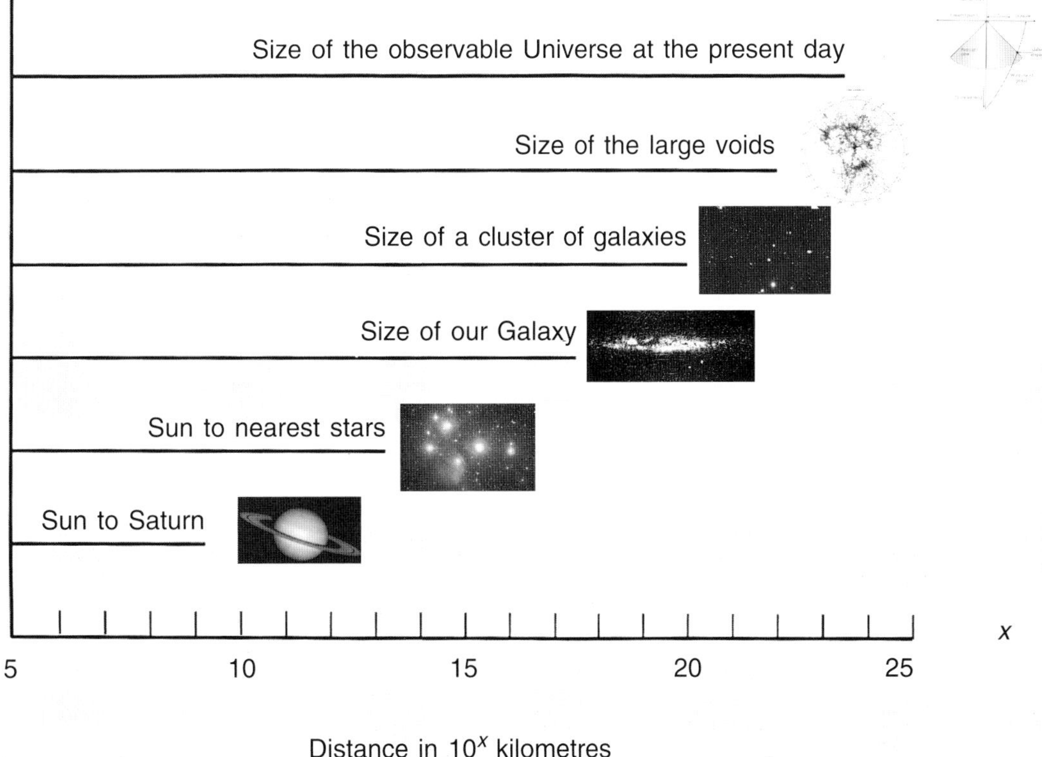

Figure 1.19. A schematic diagram showing the relative sizes and distances of different types of object in the Universe. The horizontal axis is the quantity x which is the logarithm of the distance, that is, if we write the distance in terms of powers of 10, $r = 10^x$ metres.

radiation was made by the great Scottish scientist James Clerk Maxwell in 1864. Because of the transparency of the atmosphere to light, observations in the optical waveband provide a simple and effective way of studying the Universe. We now know, however, that astronomical objects radiate electromagnetic waves over a much wider spectral range than the optical waveband and observations made with these other types of radiation provide complementary information about the nature of all classes of astronomical object.

There are two important points about waves of any sort. The first is the relation between the frequency and wavelength of the waves. Waves have a certain *frequency* which is the number of oscillations of the wave per second. It is denoted by the Greek letter nu, ν, and is measured in units of cycles per second, or Hertz (Hz), after the German scientist Heinrich Hertz, whose experiments fully confirmed Maxwell's electromagnetic theory in the period 1887–9. The waves also have a *wavelength* which is the distance between wavecrests and which is denoted by the Greek letter lambda, λ. When electromagnetic waves travel through a vacuum, the wavelength and frequency are related to the speed of light by the simple formula

$$\lambda\nu = c$$

where c is the speed of light. This formula can be easily understood because a wavecrest travels a distance λ in one period of the wave, which is just $1/\nu$. Now, the speed of light c is a fundamental constant of nature and so the shorter the wavelength, the greater the frequency of the waves. For example, a typical FM radio broadcast has a frequency of about 100 MHz and so the wavelength of these radio waves is $\lambda = c/\nu = 3 \times 10^8/10^8 = 3$ m, which explains why FM aerials are quite large.

The second point is that waves transport energy. If you have ever been knocked over by a breaking wave on the seashore, you will be well aware of the fact that waves carry energy. The fact that radio waves are used to broadcast information shows that some energy passes from the transmitter to the receiver. What is also important is that the shorter the wavelength, or the higher the frequency, the more energy is transported. Running through the optical spectrum from red through blue to ultraviolet wavelengths, we pass from lower to higher frequencies of oscillation of the electromagnetic waves or, equivalently, from longer to shorter wavelengths. As we have noted, however, the electromagnetic spectrum extends far beyond the red and ultraviolet ends of the optical spectrum.

Many of these different types of electromagnetic wave are familiar in everyday life. The lowest frequency waves used in astronomy are *radio waves* and these are of exactly the same type as the signals which are picked up by a radio receiver or a television aerial. They are also used in microwave ovens. If we move to somewhat shorter wavelengths, or higher frequencies, we come to the *infrared* waveband. Objects at roughly room temperature emit intense infrared radiation, the amount of radiation being a sensitive function of the temperature of the body. Thus, infrared cameras are used by rescue workers to find people trapped in collapsed buildings, since people are warmer than their surroundings. By night-vision, what is meant is taking pictures of the infrared radiation emitted by warm bodies, using infrared cameras. Infrared waves are also used in the remote controls for television sets and video-recorders. Going beyond the optical waveband, we come to the *ultraviolet* and *X-ray* wavebands. X-rays are familiar to everyone who has had an X-ray to find out if bones are fractured. The X-rays are very high frequency electromagnetic waves – the X-ray photographic process makes use of the fact that bones absorb X-rays more strongly than skin and muscle and so, when an X-ray plate is placed behind some part of the body, the plate is less exposed behind bones than behind the flesh and muscle. Finally, at the very highest energies, there are the gamma-rays, which are written using the Greek letter gamma, γ. γ-rays of very high energy are produced in natural radioactivity and in nuclear explosions and can be detected by Geiger counters. The key point is that these different types of radiation, which we encounter in everyday life, are also emitted by astronomical objects by natural processes and can be used to investigate all classes of object in the Universe.

One further key concept is needed to relate observations made in a particular waveband to the properties of the source of radiation. If objects are maintained at a particular temperature, they emit radiation most strongly at a particular wavelength. One of the great discoveries made towards the end of the last century was that there exists a simple relationship between the frequency at which most of the radiation of a hot body is emitted and its temperature – they are proportional to one another. This fundamental law of physics is known as *Wien's displacement law*, after Wilhelm Wien, one of the pioneers of the study of the radiation from hot bodies. It is useful to write down this formula because it is so important for astronomy. If T is the temperature of a hot body, measured in kelvin (K), then most of the radiation is emitted at a wavelength λ_{max} and frequency ν_{max} which are given by the expressions

$$\nu_{max} = 10^{11}\,T\ \text{Hz} \qquad \lambda_{max} = \frac{3000}{T}\ \mu\text{m}$$

In these relations, the temperature is measured relative to the absolute zero of temperature which is about –273 degrees Celsius. The wavelength of the radiation is measured in microns or micrometres, that is, millionths of a metre, which is written μm. What this relation tells us is that, if we observe radiation from any object at a particular wavelength, then we know roughly how hot the object has to be in order to radiate at that wavelength. These relations are displayed in Figure 1.20 in which the diagonal line stretching from the bottom left to top right represents Wien's displacement law. From this diagram, we can read off the typical temperature associated with a particular wavelength or frequency. For example, the optical waveband corresponds to wavelengths between 0.3 and 1 μm, the corresponding temperature range being 3000 to 10 000 K. To produce X-rays with a wavelength of 0.1 nm, the temperature of the object has to be about 30 million kelvin. We will return to this topic in Section 1.5. The frequency ranges of the different wavebands are shown in Figure 1.20.

We will use one other important concept concerning the properties of light. One of Einstein's great discoveries of that momentous year for physics, 1905, was that light possesses both wave and particle properties. In modern parlance, this is called the *wave–particle duality*. Although we can describe the properties of light in terms of electromagnetic waves, there is an entirely equivalent way of describing it in terms of a flux of particles called *photons*. The energy ϵ of each photon is very small and is given by the formula $\epsilon = h\nu$, where ν is the frequency of the waves and h is the fundamental constant known as *Planck's constant*, which has the value $h = 6.626 \times 10^{-34}$ J s. Photons travel at the speed of light and, according to quantum electrodynamics, they have wave properties which result in precisely all the classical effects of wave optics.

Which description is more useful in understanding the behaviour of electromagnetic radiation depends upon the waveband of observation. At optical wavelengths, modern astronomical instruments can now detect the arrival of individual optical photons at the detector element. In the ultraviolet, X-ray and γ-ray wavebands, it is generally simplest to think in terms of the arrival of fluxes of high energy photons. In the radio, millimetre and far infrared wavebands, it is normally simplest to think in terms of the arrival of a flux of electromagnetic radiation. There are no hard and fast rules about which description should be used in any waveband. In cosmology, for example, we will find that it is very important to think in terms of the number density of millimetre wave photons associated with the cosmic microwave background radiation.

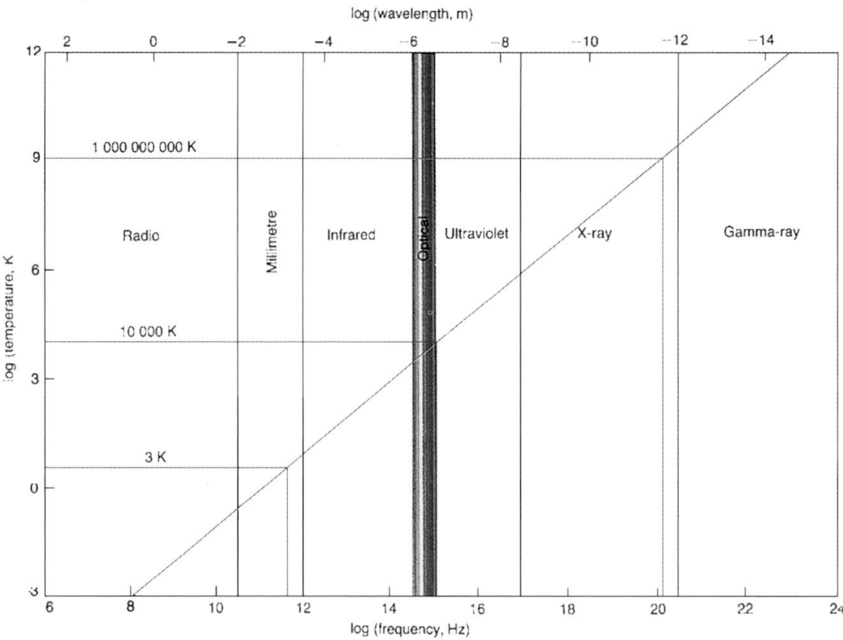

Figure 1.20. The relation between the temperature of a black body and the frequency (or wavelength) at which most of the energy is emitted. The names of the different wavebands are shown on the diagram.

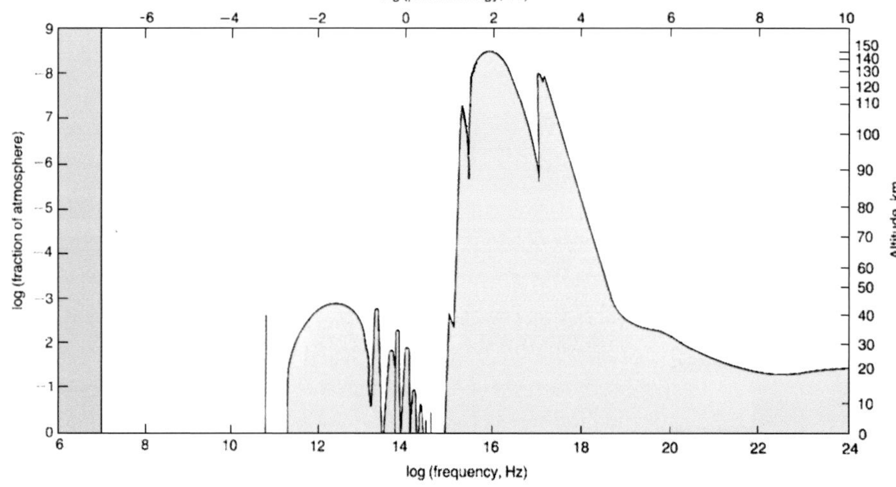

Figure 1.21. The transparency of the atmosphere for radiation of different wavelengths. The solid line shows the height above sea level at which the atmosphere becomes transparent to radiation of different wavelengths.

1.4 Observing the sky through the Earth's atmosphere

The problem for astronomy is that only certain wavebands are accessible from ground-based observatories. Figure 1.21 shows how high a telescope has to be placed above sea level in order to obtain an unobscured view of the Universe. It can be seen that optical and radio astronomy can be carried out from the ground. However, ultraviolet, X-ray and γ-ray astronomy, as well as far infrared astronomy, have to be carried out from above the Earth's atmosphere, preferably from satellites. The reason for this is the absorption of radiation by the Earth's atmosphere in these wavebands. The high energy ultraviolet, X-ray and γ-ray radiations are very dangerous and human life on Earth as we know it would not be possible without the protecting shield of the atmosphere. The infrared absorption of the atmosphere is intimately connected with the energy balance in the Earth's atmosphere and the greenhouse effect which keeps the average surface temperature of the Earth in a temperature range at which biological species can survive.

From Figure 1.21, it can be seen that an ultraviolet telescope has to be placed at least 150 km above sea level in order to obtain an unattenuated view of the ultraviolet sky and that X-ray and γ-ray telescopes can observe the sky from somewhat lower altitudes. In practice, to obtain long exposures, it is best to build the telescopes for these wavebands into satellite observatories. In the infrared waveband, it can be seen that there are some wavebands which can be observed from high altitude sites on Earth. There are certain wavelength 'windows' in the near infrared waveband at 1.2, 1.65, 2.2, 3, 5, 10 and 20 μm, through which astronomy can be conducted very successfully from the ground. Between these windows and for wavelengths between about 20 and 300 μm, the atmosphere is opaque and observations have to be made from satellites.

Although the atmosphere is transparent to optical radiation, it does have an influence upon the incoming radiation. Even on a very clear night, there are tiny fluctuations in the properties of the atmosphere above the telescope and these have the effect of slightly deflecting the paths of the incoming light rays. This is what causes the stars to twinkle. The scattering of the incoming light is known by astronomers as 'optical seeing' and it has the unfortunate effect of blurring the images of astronomical objects. This is a very serious effect for large telescopes. In theory, a 4-metre diameter optical telescope should have an angular resolution of 0.03 arcsec, that is, it should be possible to distinguish two stars which are separated by only 0.03 arcsec. In practice, the angular resolution of such telescopes is degraded to only about 1 arcsec because of astronomical seeing.

There are two solutions to this problem. One of these is to remove the effect of the atmosphere by compensating for the irregularities due to the atmosphere by incorporating special optical devices in the path of the incoming radiation. These techniques are called *adaptive optics* since they correct the incoming optical signals in real time. These are not yet standard equipment on large telescopes but the next generation of 8–10 metre optical–infrared telescopes will include such devices.

The other solution is to place the telescope above the Earth's atmosphere and this is the approach adopted by the Hubble Space Telescope, shown in Figure 1.22. Being above the atmosphere, the theoretical resolution of the 2.4-metre mirror has been achieved, following the repair mission of December 1993, which was an outstanding success. The significance of the improvement in angular resolution is illustrated in Figure 1.22 in which the image of the star cluster 30 Doradus observed with a large ground-based telescope is compared with an image of the same object observed by the Hubble Space Telescope. Two important improvements can be seen. First of all, the higher angular resolution of the Hubble Space Telescope enables much finer structure to be observed as compared with the ground-based image. Secondly, the improved resolution also makes the telescope much more sensitive for the detection of stars. Many more very faint stars can be detected in the Hubble Space Telescope image as compared with the ground-based image. It is remarkable that, for the first time, optical astronomers have obtained images at the theoretical diffraction limit of a large telescope.

1.5 The temperatures of celestial objects

The dependence of the intensity of radiation upon wavelength, or frequency, is known as the *spectrum* of the radiation. The German theoretical physicist Max Planck discovered that, at a particular temperature, the radiation spectrum of a hot body has a very simple form. This

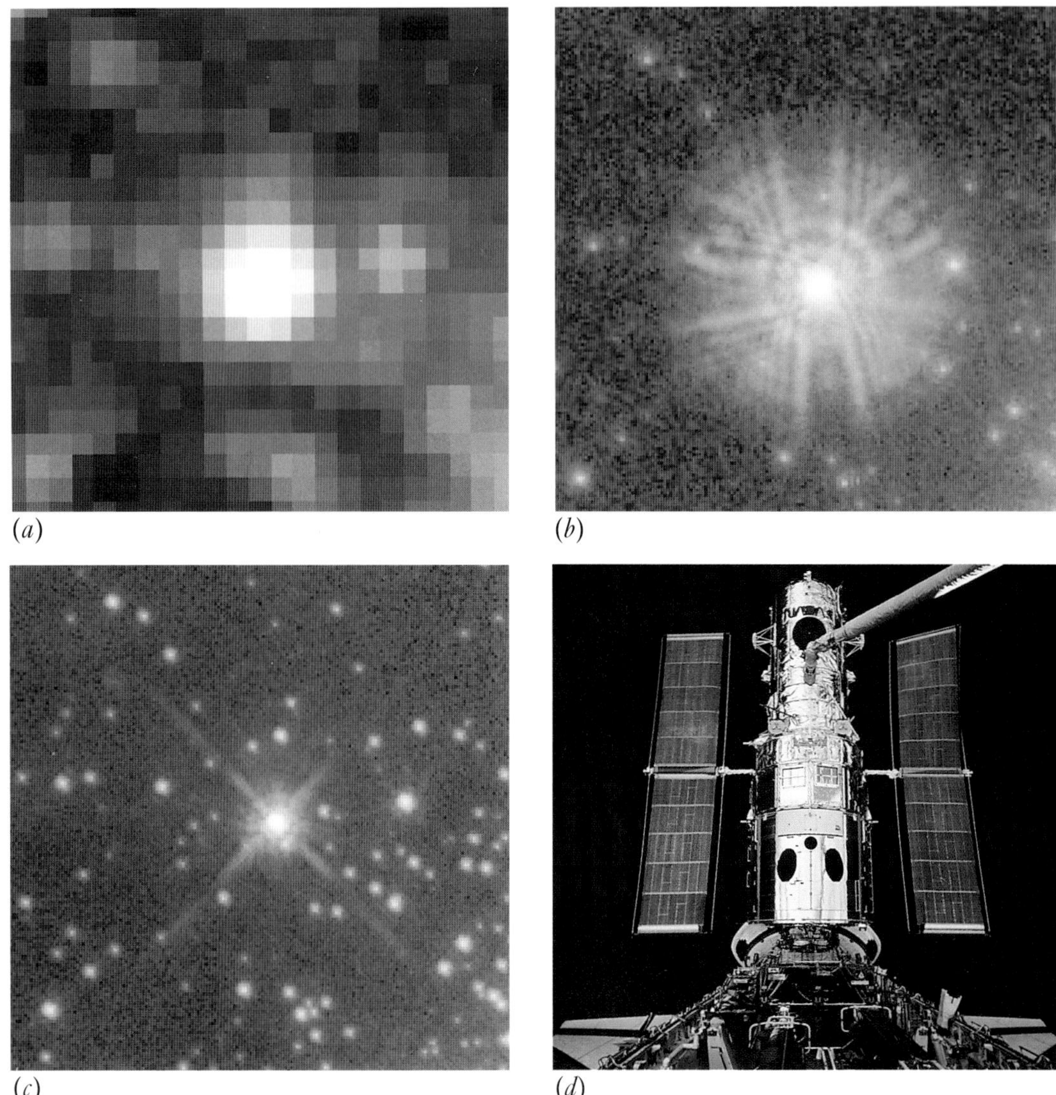

Figure 1.22. Comparison of images of the stars in the cluster 30 Doradus as observed (*a*) from the ground with an angular resolution of 0.6 arcsee; (*b*) by the Hubble Space Telescope before the repair mission of December 1993. The poor image quality, caused by spherical aberration due to incorrect polishing of the primary mirror, is apparent. (*c*) The star cluster as observed by the Hubble Space Telescope after the repair mission. Not only is the sharpness of the image improved by a factor of about 10, but also very much fainter stars can be detected. (*d*) The Hubble Space Telescope, still attached to the Space Shuttle *Endeavour*, following the highly successful repair and refurbishment mission of December 1993.

characteristic spectrum is known as the spectrum of *black-body* radiation and is observed when any hot body is maintained in thermal equilibrium with its surroundings at a single temperature. The term *thermal equilibrium* means that the body and its surroundings have been left for a very long time so that every process of absorption is balanced by the corresponding emission process. In a simple approximation, we can assume that the radiation of the stars is black-body radiation at a single temperature. The black-body energy spectra radiated by hot bodies at different temperatures in the optical waveband are shown in Figure 1.23 and have been drawn in such a way that the maximum intensity of each curve is the same, although the temperatures are different. The frequency at which the maximum intensity occurs is given by Wien's displacement law, which was described in Section 1.3. The diagram shows the expected thermal distributions of radiation for bodies at temperatures of 3000, 5700, 12 000 and 20 000 K. Notice the characteristic shape of the energy spectra – radiation is emitted at a wide range of wavelengths but most of it is emitted at a wavelength, or frequency, close to the maximum intensity value.

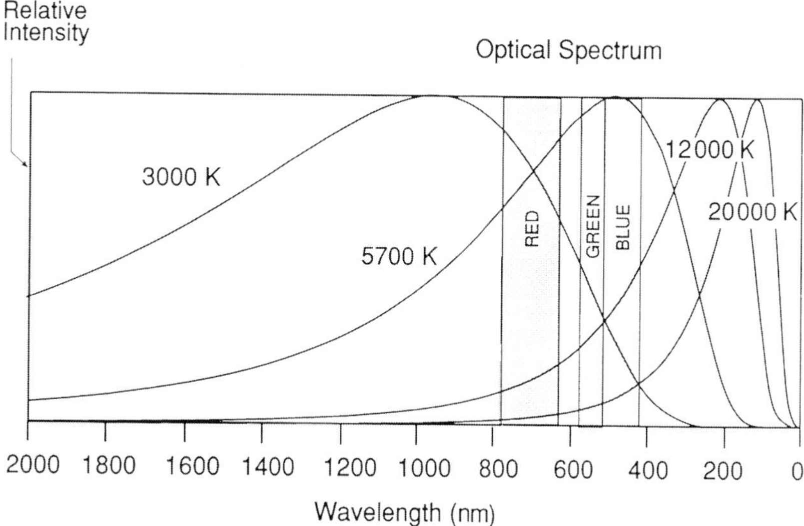

Figure 1.23. Illustrating the radiation spectra expected of hot bodies at temperatures of 3000, 5700, 12 000 and 20 000 K. The lines show the intensity spectra of black-body radiation and the wavelength range shown includes the optical waveband. The red, green and blue regions of the optical spectrum are indicated.

As the temperature increases, the maximum of the radiation spectrum moves to shorter and shorter wavelengths according to Wien's displacement law, $\lambda_{max} = 3000/T$ µm. The red, green and blue wavebands of the optical spectrum have been indicated, because these are the familiar colours in the optical spectrum to which our eyes are sensitive. The colours of the stars can be estimated by noting in which of these wavebands the greatest intensity occurs. It can be seen that, in cool stars at 3000 K, the maximum intensity in the optical region of the spectrum occurs in the red wavelength interval and so cool stars are red. At the other extreme, it can be seen that hot stars with temperatures 12 000 and 20 000 K have maximum intensities in the blue region of the optical spectrum and so are blue stars. Our Sun has a surface temperture of about 5700 K. The spectrum of a black body at this temperature is also shown, and it can be seen that there are roughly equal intensities of red, green and blue light, which results in a more or less white colour with a tinge of yellow, exactly the colour of the Sun.

This is a very important result because it tells us that when we make observations in different wavebands, either the radio, infrared, optical, ultraviolet, X-ray or γ-ray wavebands, we obtain different *temperature pictures* of the Universe. In modern astronomy, we can observe objects with temperatures which are within a few degrees of absolute zero by observing in the centimetre and millimetre regions of the radio waveband right up to those with temperatures as high as a billion kelvin and more, which can be observed with γ-ray telescopes.

Let us demonstrate how this works for a particular object. About 250 years ago, a star exploded in the constellation of Cassiopeia. In the optical waveband, straggly optical filaments are observed moving away from the site of the explosion (Figure 1.24). These filaments are cooling by radiating away their energy in the optical waveband, because their temperatures lie in the correct temperature range, about 10 000 K, to be observed at optical wavelengths. The same object is much more spectacular when observed in the X-ray waveband. Figure 1.25 shows the same region of sky but now as observed by the Einstein X-ray satellite. These observations show that the remnant contains a sphere of very hot gas. To be observable in the X-ray waveband, the typical temperature of the gas filling the remnant has to be about 10 million kelvin. Thus, the combination of optical and X-ray observations shows that there

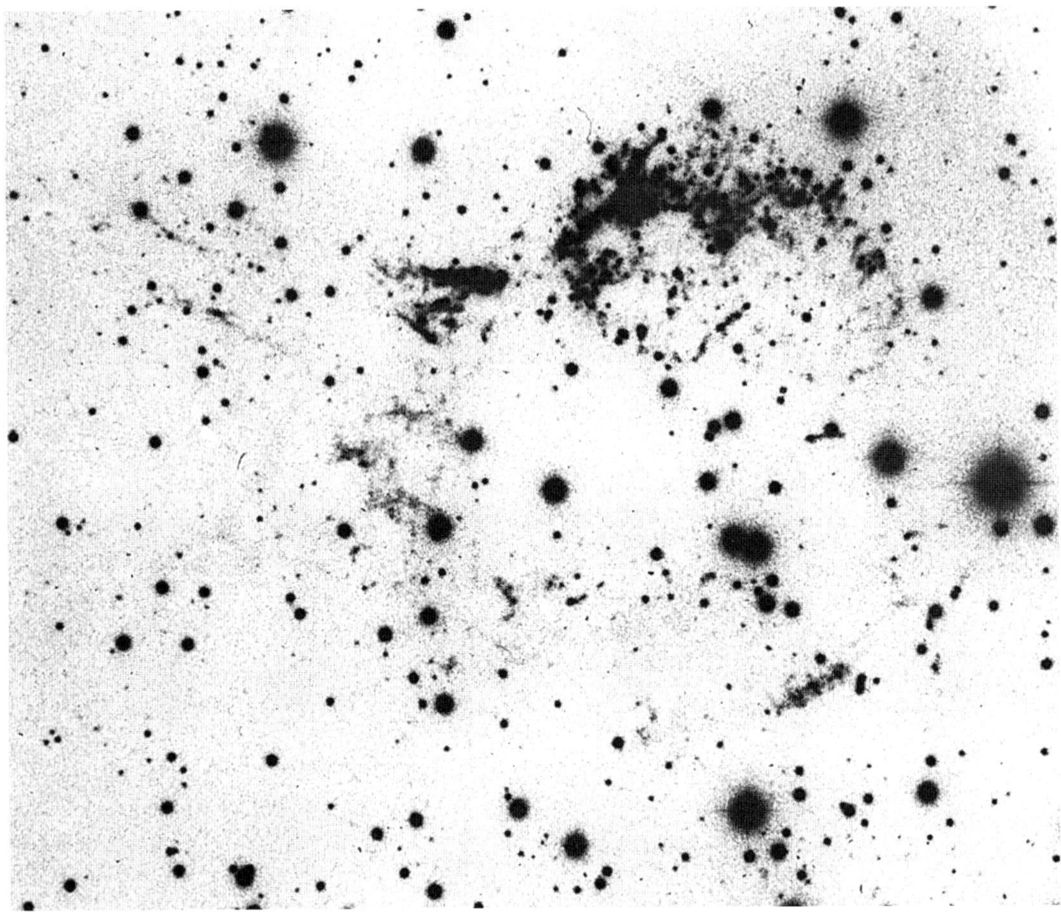

Figure 1.24. An optical photograph of the supernova remnant Cassiopeia A. The wispy filaments are moving away from the site of the exploding star at a speed of about 10 000 km s^{-1}.

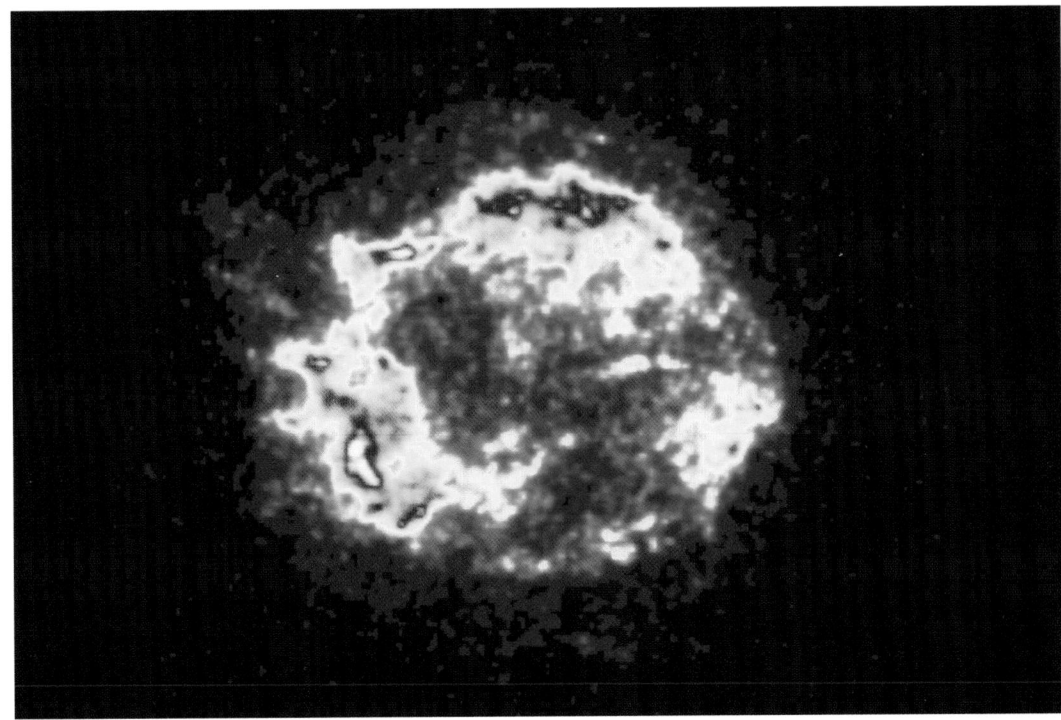

Figure 1.25. An X-ray image of the supernova remnant Cassiopeia A. The gas has been heated to a temperature of about 10 000 000 K by the explosion of the star.

is very hot gas inside the supernova remnant, which is pushing back the surrounding interstellar gas which is heated and compressed by the supersonic expansion of the hot gas.

Another very good example concerns studies of one of the regions closest to the Earth, in which massive stars are being born. In the optical waveband, the Orion Nebula is one of the most striking objects in the sky with its beautiful optical filamentary structures (Figure 2.14). All this gas is at a temperature of about 10 000 K and is heated by the young blue massive stars, the four Trapezium stars, in the centre of the Nebula. When this region is observed in the infrared and millimetre regions of the spectrum, we obtain a completely different picture. We see that the Orion Nebula is surrounded by a vast cloud of cold molecular gas and it is out of this cooler material that even younger stars are forming (Figure 2.19). These enormous giant molecular clouds, observed at centimetre and millimetre wavelengths, are the nurseries in which stars are formed. The Orion Nebula is simply one of the densest parts of the cold giant molecular cloud which has collapsed under gravity and formed the four bright luminous stars which illuminate the nebula. We will return to these beautiful observations in Chapter 2.

These two examples illustrate how the different wavebands provide complementary information, which leads to a much more complete picture of the processes responsible for the formation and evolution of these objects.

1.6 The Universe as observed in different wavebands

Let us look at the whole sky again, but now as observed in all the different wavebands discussed above. First of all, let us recall Figure 1.7 which shows the *optical* picture of the whole sky, with which we are familiar. As explained in Section 1.2, we can represent the whole sky on a single sheet of paper by adopting the Aitoff projection shown in Figure 1.7. This optical picture of the sky includes all the stars brighter than about 10th magnitude, as well as all the nebulae seen in the plane of the Milky Way. Most of the light we see in this picture is due to gas and stars, radiating at temperatures between about 3 000 to 10 000 K. There are very good astrophysical reasons why the stars radiate with surface temperatures which lie in this temperature range, as we will discuss in Chapter 2. In Figure 1.7, our two dwarf neighbouring galaxies, the Large and Small Magellanic Clouds, can be observed to the lower right of the Galactic plane.

Let us now look at a 'hotter' image of the sky in the same Aitoff projection. Figure 1.26(a) is a map of the sky in the *X-ray waveband* made by the first High Energy Astrophysical Observatory satellite (HEAO-1) and it is clearly a rather different picture. There are some sources of X-ray emission concentrated towards the Galactic plane but they are far fewer than in the optical picture. The reason for this is simple. Objects which radiate X-rays must have temperatures of about 10 000 000 K and these are very rare objects in our Galaxy. There are two broad classes of Galactic X-ray source. Many of them are associated with binary star systems in which a very compact star radiates at a very high temperature because of mass falling onto its surface from the companion star. We will discuss these binary X-ray sources in Chapter 3, in the context of the physics of active galactic nuclei and quasars. The other broad class of Galactic X-ray source consists of the remnants of exploding stars, the supernova remnants such as Cassiopeia A (Figures 1.24 and 1.25), and the dead stars formed in these explosions.

Away from the Galactic plane, it can be seen that there is a large population of objects, many more than were present in the corresponding optical picture. These X-ray sources are mostly associated with active galaxies and quasars at very great distances. The reason they

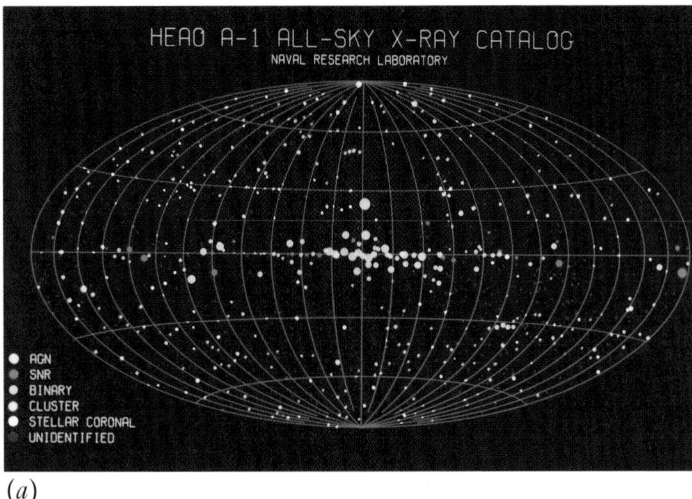

(*a*)

(*b*)

Figure 1.26. (*a*) An X-ray image of the whole sky in the same projection as Figure 1.7 obtained from the HEAO-1 satellite. (*b*) The German ROSAT X-ray observatory with UK and US participation, which was launched in 1990 and which has made the most complete survey of the sky at X-ray energies of about 1 keV.

Figure 1.27. (*a*) The γ-ray map of the Milky Way as observed by the Compton Gamma-Ray Observatory (*b*) The Compton Gamma-Ray Observatory, which was launched in April 1991 and which completed the first all-sky γ-ray survey, illustrated in (*a*).

appear in such numbers on the sky is that these objects are very luminous X-ray emitters. There are also a few clusters of galaxies which contain large quantities of very hot gas in the space between the galaxies.

We can study even higher temperatures from the γ-ray map of the sky (Figure 1.27(*a*)). This image of the whole sky was made by the Compton Gamma-Ray Observatory of NASA (Figure 1.27(*b*)), named after the physicist Arthur Holly Compton whose experiments finally convinced everyone that light consists of a flux of photons. This map was obtained by the energetic γ-ray telescope, which is sensitive to γ-rays with energies greater than 100 MeV. It can be seen that the bulk of the γ-radiation is confined to a thin layer in the plane of our Galaxy. The γ-ray emission must be associated with a very high temperature gas and, in fact, is associated with very high energy particles which travel at speeds close to the speed of light in the space between the stars. These particles are very similar to the very high energy *cosmic ray* particles, which can be detected by particle detectors in satellites at the top of the Earth's atmosphere. The γ-rays are due to collisions between these very high energy particles and the cold interstellar gas. The importance of these observations is that they show that there are high energy particles, as well as ordinary gas, in what is called the *interstellar medium*. There is a close correlation between regions in which there is known to be enhanced density of the interstellar gas and γ-ray features on the

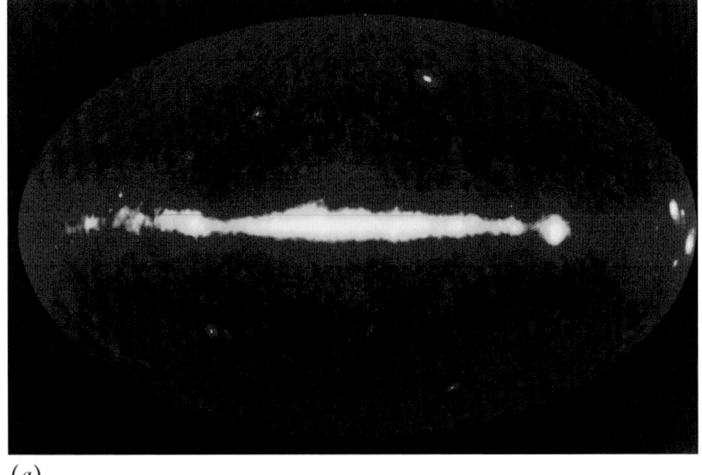

(*a*)

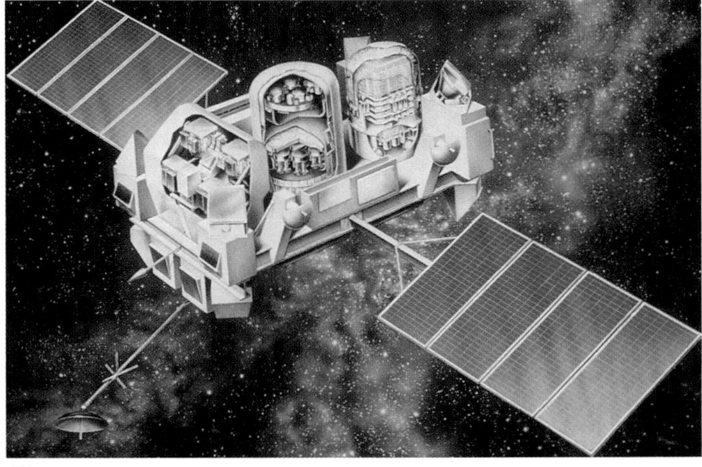

(*b*)

map, as can be observed by comparing Figure 1.27(*a*) with Figure 1.28(*a*), which, as we will discuss, is primarily associated with the distribution of interstellar dust.

In addition to intense emission from the plane of the Galaxy, a few γ-ray sources have been observed at high Galactic latitudes. These have been found to be associated with the most violent sources of energy we know of anywhere in the Universe. These objects are the extremely violently variable radio quasars, many of which exhibit superluminal motions. 24 such objects were detected in the first year of observation and they have posed intriguing problems for high energy astrophysicists.

Let us now consider lower temperatures than those of the stars observed in the optical waveband. To achieve this, observations have to be made in the *infrared* region of the spectrum and Figure 1.28(*a*) shows a map of the whole sky made at wavelengths of 60 to 100 μm by the Infrared Astronomical Satellite (IRAS). These wavelengths correspond to temperatures of only about 60–100 K, that is, very much less than room temperature, in fact, to about –170 to –210 degrees Celsius. Once again, there is a narrow distribution of intense radiation lying along the Galactic plane. Although this matter is very cold, the radiation is very intense. It is due to the radiation of cool interstellar dust lying in the plane of the Galaxy. Why is the dust at this low temperature? The reason is that, although the dust is not very hot, it must be heated or else it would cool down to a very low temperature by radiating away all its internal energy. What is happening is that the dust grains are heated to a temperature of about 60 to 100 K by the radiation of stars, particularly those in the process of formation and those which are dying. The dust grains then radiate, more or less like little black bodies, at the temperature to which they have been heated. This process is believed to be responsible for most of the far infrared radiation of our Galaxy. Infrared observations of cool dust and molecular gas are among the most important tools for studying how stars are formed (Chapter 2).

This process of the absorption of optical and ultraviolet radiation by dust followed by its reradiation at far infrared wavelengths is beautifully illustrated by the maps of the sky at different wavelengths. Much of the optical picture of the Galaxy (Figure 1.7) is obscured by

Figure 1.28. (*a*) The whole sky as observed by the Infrared Astronomical Satellite (IRAS) in the wavebands 60 to 100 μm. (*b*) The Infrared Astronomical Satellite (IRAS) which completed the first far infrared surveys of the sky in the period 1981 to 1983. The telescope was cooled to liquid helium temperatures in order to reduce its emissivity to a very low value in the far infrared wavebands.

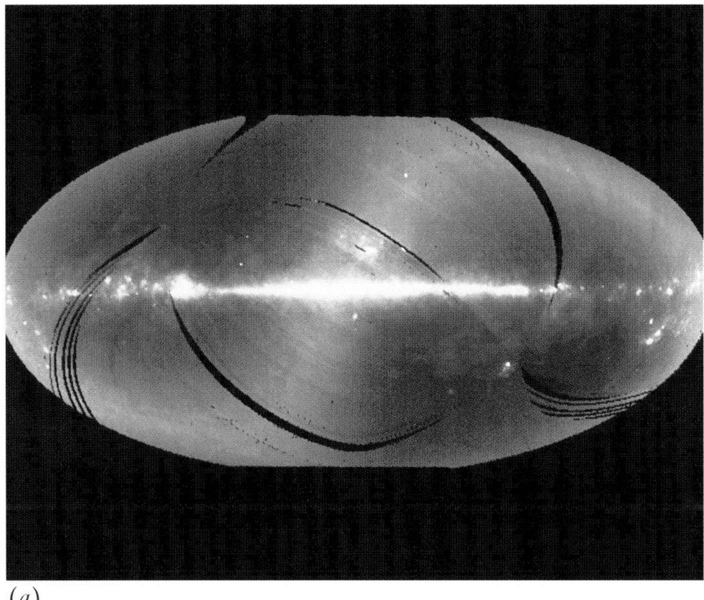

(*a*)

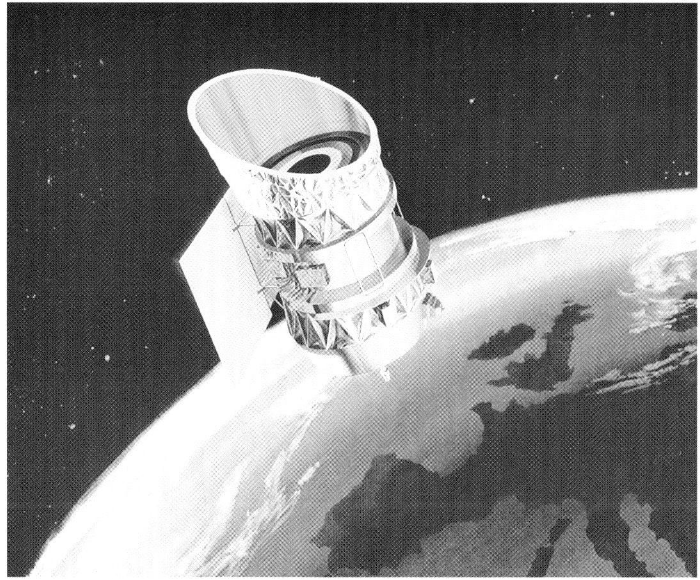

(*b*)

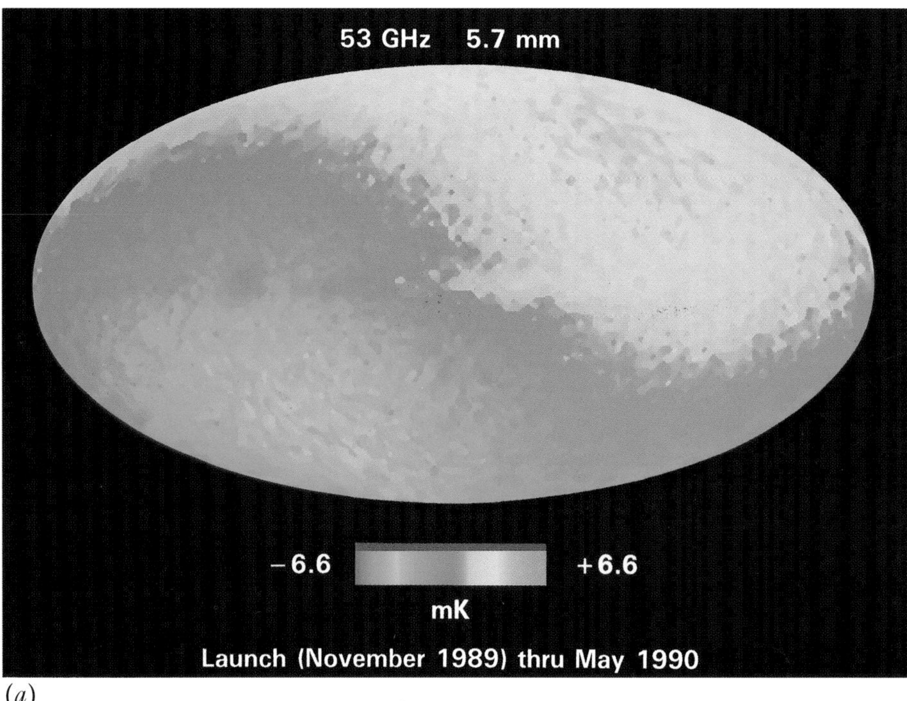

(*a*)

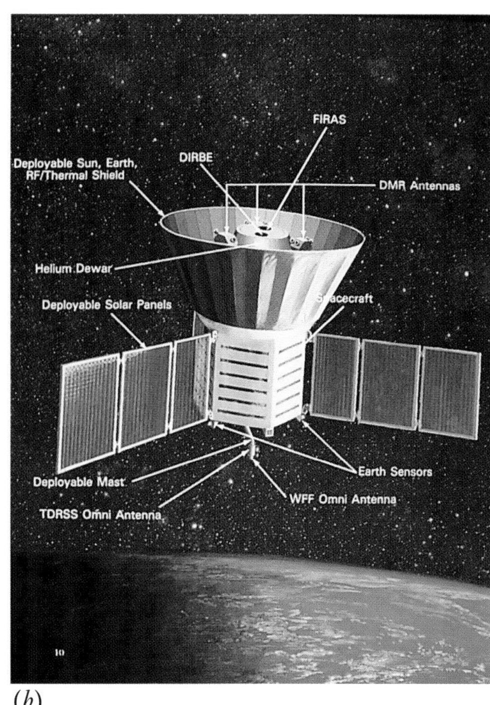

(*b*)

Figure 1.29. (*a*) The map of the whole sky as observed in the millimetre waveband at a wavelength of about 5.7 mm by the COBE satellite. The image is dominated by the dipole component of the microwave background radiation. (*b*) The Cosmic Background Explorer (COBE), which was launched in November 1989. In addition to the background radiation in the millimetre and sub-millimetre regions of the spectrum, observations have been made throughout the near and far infrared wavebands as well. An example of one of these images is shown in Figure 1.8.

interstellar dust. As we mentioned in discussing that image, the dust becomes transparent at infrared wavelengths. This is why we were able to obtain such a beautiful image of the disc and bulge of our Galaxy from the near infrared observations made with the COBE satellite (Figure 1.8). In exactly the same way, Figure 1.28(*a*) is an unobscured picture of the whole Galaxy at far infrared wavelengths. Now, however, the dust is no longer absorbing the radiation of the stars but is reradiating all the absorbed energy in the far infrared waveband.

As we map the sky at longer and longer wavelengths, we expect to observe colder and colder objects and Figure 1.29(*a*) is an image of the whole sky at a *millimetre* wavelength of only 5.7 mm made by the Cosmic Background Explorer satellite (COBE) of NASA (Figure 1.29(*b*)). This map corresponds to temperatures only about 3 degrees above absolute zero. This picture differs dramatically from all the others we have looked at so far. In the other pictures, we saw a clearly defined Milky Way and there was sometimes evidence for sources lying in directions away from it. In Figure 1.29(*a*), there is a vestige of the Milky Way running along the centre of the diagram, indicating that there is some very cool gas present in the plane of the Galaxy, but the picture is dominated by intense radiation which is brighter towards the top right of the map and fainter towards the bottom left. This picture is, in fact, not of the total intensity of the radiation but only of tiny intensity deviations, at the level of one part in 10 000, from a uniform distribution at a temperature 2.725 K. This is a map of the famous cosmic microwave background radiation, the cooled remnant of the very hot early stages of the Big Bang, which will be discussed in much more detail in Chapters 4 and 5. The reason that one half of the sky is slightly hotter than it is in the opposite direction is entirely due to the motion of the Earth through the frame of reference in which the radiation would be perfectly uniform over the whole sky. This is no more than the result of the Doppler effect which, because of the motion of the Earth through the background radiation,

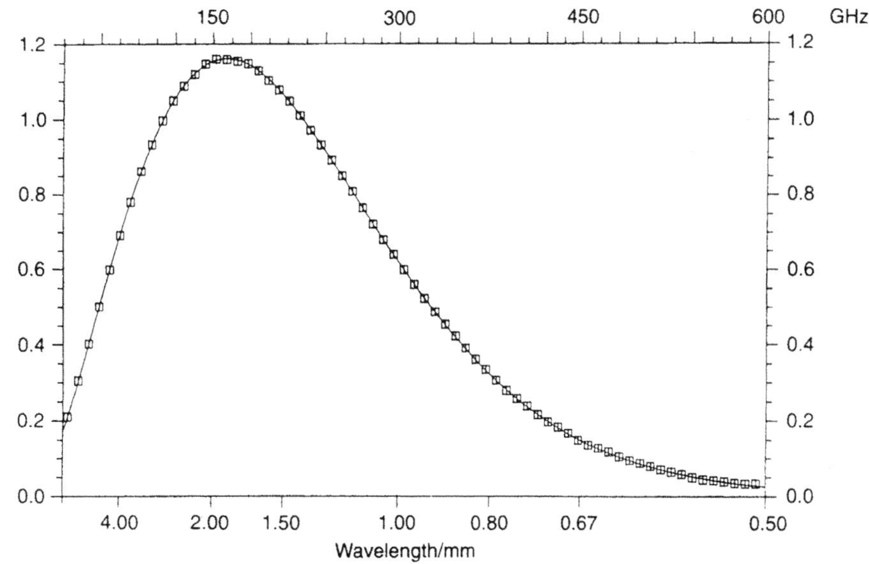

Figure 1.30. The spectrum of the microwave background radiation as measured by the COBE experiment. The spectrum is that of a black body at a temperature of 2.725 K. It is now known that the spectrum has the form of a perfect black body to an accuracy of 0.03% of the maximum intensity in the wavelength range 0.5 to 2.5 mm.

makes it appear to be more intense in the direction of motion. This effect has to be removed if we are to study the distribution of the radiation on a large scale. The radiation then turns out to have exactly the same intensity in all directions to a precision of about one part in 100 000. At that level, tiny variations in intensity of cosmological origin have been observed. These are crucial observations for cosmology and we will take up this story in Chapters 4 and 5.

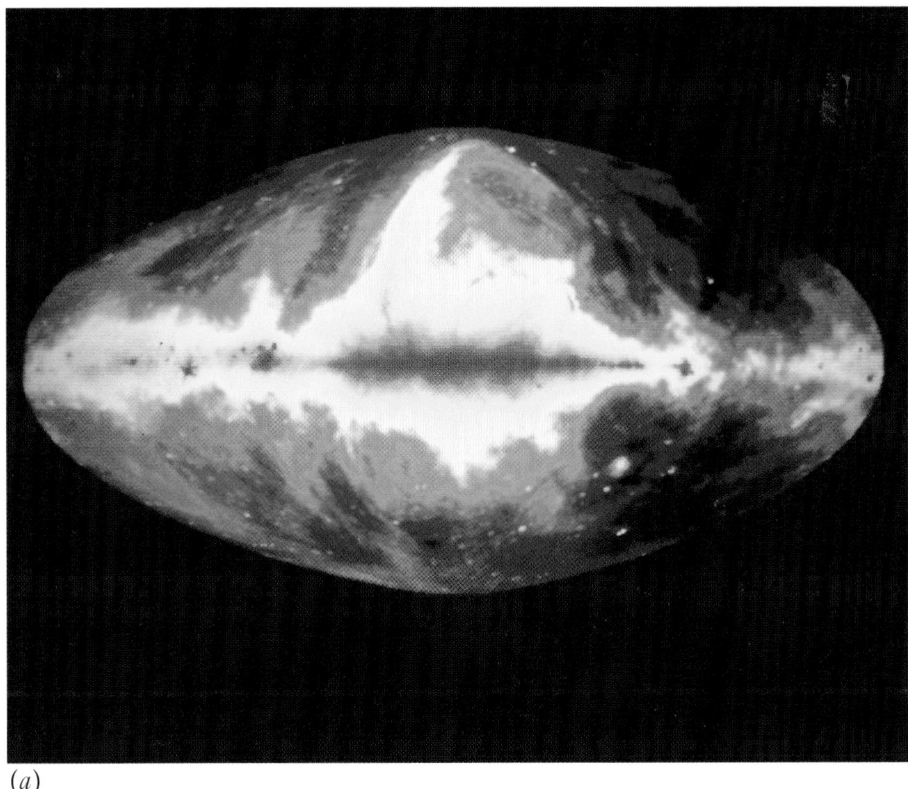

(*a*)

(*b*)

Figure 1.31. (*a*) The radio emission from the Galaxy as measured at a wavelength of 73 cm (408 MHz) by the radio astronomers at the Max Planck Institute for Radio Astronomy at Bonn. (*b*) The Effelsberg 100-metre radio telescope of the Max Planck Institute, which was used to map the northern celestial hemisphere shown in (*a*).

One of the most remarkable features of the cosmic microwave background radiation is that it has a perfect black body spectrum, one of the most amazing observations of modern cosmology, again made by the Cosmic Background Explorer (Figure 1.30). This is the most perfect black body spectrum we know of anywhere in nature. We will explain why it has this form in Chapter 5.

If the sky is observed at even longer wavelengths, in the *radio* waveband, we might expect to detect the coldest material of all but, in fact, this is not the case. The radio map of the sky shown in Figure 1.31(*a*) was made by the radio astronomers at the Max Planck Institute for Radio Astronomy at Bonn, Germany, at the long radio wavelength of 73 cm (408 MHz) (Figure 1.31(*b*)). Again the Galactic plane appears prominently but now there are spurs and plumes of radio emission bubbling out of the plane. It turns out that the radio emission is associated with the radiation of ultra-high energy electrons gyrating in the weak magnetic field which is present in interstellar space. This radiation is known as *synchrotron radiation* and is one of the most important processes for studying high energy phenomena wherever they occur in the Universe. The radio map of the Galaxy is convincing evidence that there are high energy electrons and magnetic fields present throughout the plane of the Galaxy. Thus, whereas the γ-ray map of the Galaxy (Figure 1.27(*a*)) shows where high energy protons and nuclei are found in the Galaxy, the radio map indicates where the high energy electrons are located. Comparison of the maps of the Galactic plane in Figures 1.27(*a*) and 1.31(*a*) shows that they are remarkably similar, which is an important result. The radio emission from the Galaxy is similar in its properties to the radiation which is observed from exotic objects such as the quasars and active galactic nuclei. The synchrotron radiation from these objects provides one of the most important tools for studying the astrophysics of active galactic nuclei (Chapter 3).

1.7 The multi-wavelength Universe

In this chapter, we have presented a broad-brush picture of the Universe and its contents. The most important conclusion has been that we have to observe the Universe in all the wavebands accessible to us, if we are to obtain a complete picture of the remarkable diversity of objects present in the Universe and the physical processes responsible for their origin and evolution. We have to account for the origin and evolution of stars, galaxies and active galactic nuclei, as well as the presence of interstellar dust and gas in all its phases, very hot gas as indicated by its X-ray emission and very cold gas found in regions where stars are formed. We also need to explain the origin of the high energy particles and magnetic fields observed in the interstellar medium in galaxies and, even more tantalisingly, in active galaxies.

Astronomical objects provide us with environments which are quite different from those which can be studied in terrestrial laboratories. One of the most exciting aspects of these studies is that occasionally we have to introduce quite new physical concepts, which can only be studied in astronomical environments. A classic example of this is the study of the black holes which are needed to understand the physics of active galactic nuclei and certain binary X-ray sources. The quest is therefore twofold. We aim to understand the physical processes responsible for the origin and evolution of all the contents the Universe and, in the process, to obtain a deeper understanding of the laws of physics themselves. In the next four chapters, our objective is to put all these diverse observations into context and create a convincing picture for the origin and evolution of stars, galaxies, active galactic nuclei and the Universe itself. In all these studies, we will encounter difficult, unsolved problems.

2 The birth of the stars and the great cosmic cycle

2.1 Powering the Sun and the stars

The subject of this chapter is the stars – what they are, how they evolve and how they were formed. Most of the visible mass of our Galaxy is in the form of stars and so there must exist efficient ways of condensing what started out as very diffuse gas into the very compact objects we call stars. Before addressing that problem, let us study what we know about the internal structure of the stars, in particular, our own Sun, and the physical processes taking place in their interiors. Then, we will describe how different types of star evolve, in the process synthesising the chemical elements.

First of all, let us tackle the problem of what it is that makes the stars shine. This was one of the great problems of 19th century astronomy. The most popular idea at that time, which was formulated by Lord Kelvin and Hermann von Helmholtz, was that the source of energy was the slow contraction of the Sun. In this picture, as the Sun slowly contracts, its gravitational energy increases and so the Sun can feed off this energy in order to maintain its luminosity. The problem with this theory was that the Sun has only enough gravitational energy to keep it radiating at its present luminosity for about 10 million years. This time-scale, known as the *Kelvin–Helmholtz time-scale*, is in serious conflict with the age of our Solar System, which is known to be about 4.6 billion years from radioactive dating.

The problem of the origin of the Sun's energy was only solved once Einstein had discovered, in 1905, the equivalence of mass and energy, perhaps the most famous equation in physics, $E = mc^2$. This relation tells us how much energy, E, is released if a mass m is completely converted into energy, or equivalently, how much mass is associated with a given amount of energy. In 1920, precise values for the masses of atomic nuclei were being measured by F. W. Aston at the Cavendish Laboratory in Cambridge. Arthur Eddington noted in his address to the meeting of the British Association for the Advancement of Science in Cardiff in the same year that, if four hydrogen nuclei, or protons, come together to form a helium nucleus, the mass of the helium nucleus is slightly less than the sum of the masses of the four protons. The 'missing mass' must be released as energy, according to Einstein's formula. Eddington showed that, if this nuclear reaction could take place, about 1/120 of the rest mass energy of each proton could be released to power the Sun. It was a number of years before this proposal could be put on a firm physical foundation. Once quantum mechanics had been properly formulated in the mid-1920s, it became possible to make the first predictions of the reaction rates. By the end of the 1930s, the rates of the different nuclear reactions which convert hydrogen into helium were understood and Eddington's hypothesis was found to be absolutely correct. The energy source of the Sun is the nuclear burning of hydrogen into helium and we now know that there is sufficient hydrogen fuel in the Sun for it to radiate at its present luminosity for about 10 billion years.

The understanding of the physics of stars like the Sun is one of the great triumphs of modern astrophysics and studies of their structure and evolution are nowadays among the most precise of the astrophysical sciences. Let us first demonstrate just how well the theory can account for the structure and evolution of the stars before tackling the more difficult problem of how they were formed.

2.2 The Sun – a testbed for the physics of the stars

The process of nuclear energy generation in the Sun is quite straightforward. Most of the matter in the Universe is in the form of hydrogen, the lightest of all the chemical elements. Each atom of hydrogen consists of a proton, which has a single positive electric charge, surrounded by the negative charge cloud associated with a single electron. Heavier elements such as helium, carbon and oxygen have nuclei which are heavier than those of hydrogen and are composed of particles called nucleons, which are either protons or their neutral twins, the neutrons. These atoms are shown schematically in Figure 2.1. Since the protons all have the same positive electric charge, each atom is surrounded by a cloud of electrons with the same total negative electric charge so that overall the atom is electrically neutral.

It is a general rule that, in all natural processes, systems try to reach their lowest energy state. The fact that the nucleons are firmly bound together in atomic nuclei means that it must be energetically favourable for the nucleons to combine together to form heavier nuclei rather than to remain as isolated protons and neutrons. As was noted by Eddington, if a helium

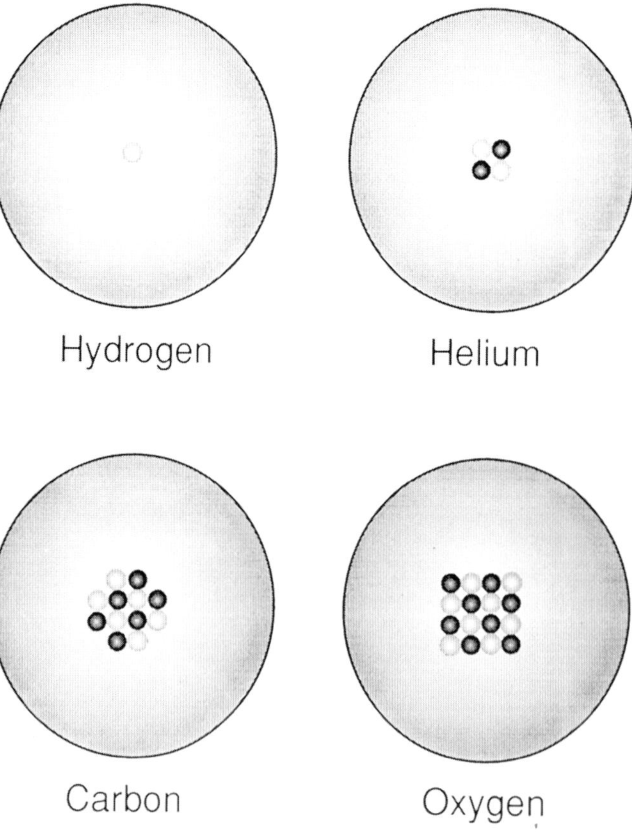

Figure 2.1. Sketches of the atoms of hydrogen, helium, carbon and oxygen. Hydrogen consists of a single proton, which has a single positive charge, surrounded by the charge cloud associated with a single electron, which has a single negative charge. The helium nucleus consists of two protons and two neutrons. The neutron has no electric charge, and so the charge cloud is associated with two negatively charged electrons. In the same way, the nucleus of carbon contains six protons and six neutrons and oxygen has eight protons and eight neutrons.

nucleus is formed by combining four protons together, some energy is left over, indicating that the nucleus has reached a lower energy state. This reaction, the synthesis of helium out of hydrogen, is not only the source of energy in stars like the Sun, but is also the first step in the synthesis of heavier elements such as carbon and oxygen.

The region in which these nuclear reactions take place occupies roughly the central 10% of the Sun by radius. The temperature in the centre of the Sun is about 16 000 000 K and this is sufficiently high for the burning of hydrogen into helium to take place. Energy is transported from the central regions to the outer envelope by radiation and this process is responsible for maintaining the Sun's internal thermal energy. Maintaining the internal energy of the Sun is essential since this is the source of the internal pressure which is responsible for holding the Sun up against the attractive force of gravity. It is a straightforward calculation to show that the Sun must be in a state of almost perfect balance between the inward pull of gravity and the outward pressure of the hot stellar material (Figure 2.2). This state is what we mean by *normal stars* – they are regions in the Universe in which the gas density and the temperature are so great that nuclear reactions take place and the energy released provides the pressure support for the star.

How do we know that this is a correct description of the internal structure of the Sun? After all, with optical telescopes, only the surface of the Sun can be observed, in fact, only the very thin surface layer from which light is emitted. Although the central regions of the Sun are very hot, the temperature decreases outwards and the surface region, from which the

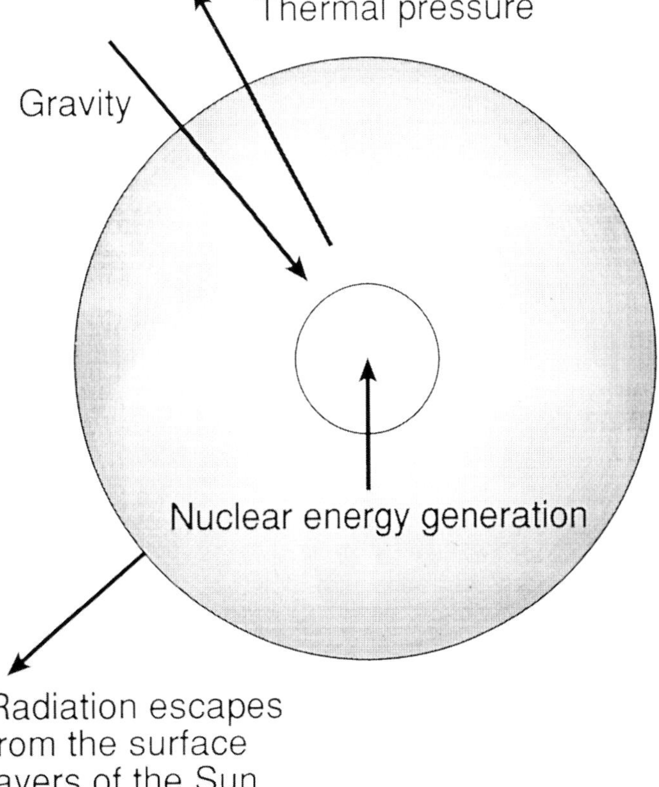

Figure 2.2. A schematic diagram showing the internal structure of the Sun. The Sun is held up by internal pressure gradients against the attractive force of gravity. The source of energy to maintain the internal pressure of the Sun is the nuclear burning of hydrogen to helium in its centre.

light is emitted, is at a temperature of only about 5700 K. The traditional method of determining the internal structure of the Sun and the stars is to infer these properties indirectly from observations of their surface properties. High precision spectroscopy continues to be one of the most important ways of undertaking these studies but there are now two new methods of probing the interior structure of the Sun.

The first of these techniques is now more than 25 years old and involves searching for certain types of particles which are created in the nuclear reactions in the centre of the Sun and which are not absorbed by the material of the Sun. These are the particles known as *neutrinos*. What are neutrinos? They are particles involved in a particular type of nuclear reaction, known as the weak interaction, and their existence was predicted theoretically by Wolfgang Pauli in 1931 to account for certain features of these interactions. The neutrinos are very difficult to detect because they hardly interact with matter at all. They were only detected experimentally in 1956 by Clyde Cowan and Fred Reines, who placed huge neutrino detectors underneath a nuclear reactor. Neutrinos are liberated in quite enormous numbers by the nuclear reactions taking place in the centre of the Sun. Because they interact so weakly with matter, they escape more or less directly from their point of origin without being absorbed by the material of the Sun. Each second there are roughly 10^{15} neutrinos passing through each square metre of the Earth's surface, but they interact so weakly with matter that we are not aware of their presence.

It is of central importance to detect this flux of solar neutrinos, because they provide a direct test of our models for the nuclear physics and astrophysics of the nuclear burning regions in the centre of the Sun. Let us look at the processes of neutrino production in a little more detail. In the Sun, the formation of helium from four protons takes place through a sequence of reactions known as the proton–proton, or p–p, chain, which is illustrated in Figure 2.3(*a*). First, two protons combine to form a deuterium nucleus, or a deuteron D, which consists of a neutron and a proton – this is a very rare weak interaction and is the crucial first step in the p–p chain. In forming a deuteron, a positive electron and an antineutrino are released. Notice that it is an antineutrino rather than a neutrino which is emitted in this reaction. All particles have their opposite counterparts in nature, in which all their fundamental properties have the opposite sign – these forms of matter are called antimatter. For example, the positron is the antiparticle of the electron and has a positive rather than a negative electric charge. In the same way, the antineutrino is the antiparticle associated with the neutrino and it happens to be one of these particles which is released in the formation of a deuteron. We will encounter antimatter again in Chapter 5 in our studies of the physics of the early Universe. The maximum energy of the antineutrino released in this process is only about 0.4 MeV (megaelectronvolts). The next step in the chain involves adding another proton to the deuteron to make an isotope of helium which has three nucleons, two protons and one neutron – it is known as helium-3, ^{3}He. Finally, two helium-3 nuclei come together to form helium-4, ^{4}He, ejecting the two unneeded protons.

In 1955, Raymond Davis suggested that solar neutrinos could be detected by the nuclear reactions which they induce in the nuclei of chlorine atoms. If a neutrino collides with one of these nuclei, it converts it into a radioactive argon nucleus and so the number of neutrinos can be measured by the number of radioactive argon nuclei created in a chlorine detector. The problem with this idea is that the neutrinos have to have an energy of at least 0.814 MeV

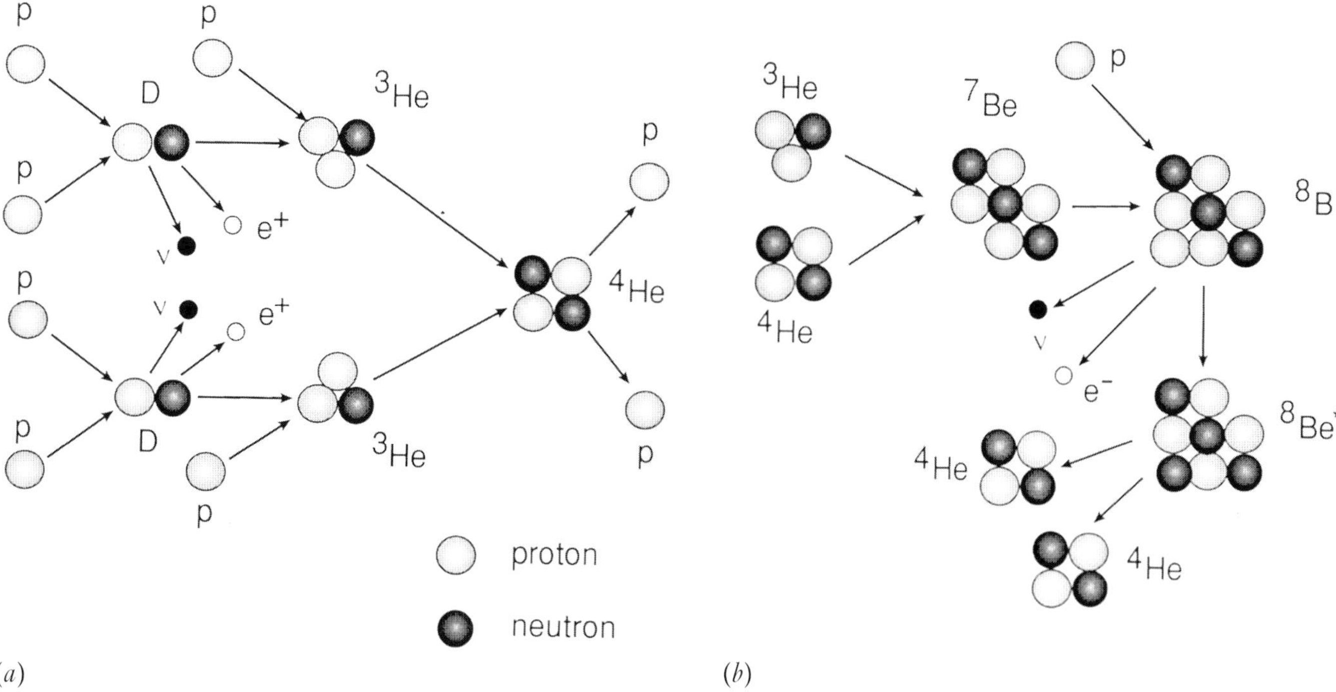

Figure 2.3. (*a*) Illustrating the proton–proton, or p–p, chain for the formation of helium from protons. Notice the low energy antineutrino emitted in the crucial first step in the chain, the formation of deuterium. (*b*) The side-chain of the main p–p chain responsible for the production of energetic neutrinos from the decay of ^{8}B.

before the chlorine nucleus can be converted into an argon nucleus and so the neutrinos produced during deuterium formation could not be detected by this means. In 1958, however, Alistair Cameron and William Fowler pointed out that there is a side reaction of the main p–p chain which results in the emission of much higher energy neutrinos (Figure 2.3(*b*)). Occasionally, beryllium-7, ^{7}Be, which has four protons and three neutrons, is formed by the collision of helium-3 and helium-4 nuclei. If a proton is added to the beryllium-7 nucleus, boron-8, ^{8}B, is produced, which decays into the unstable nucleus beryllium-8, ^{8}Be, with the emission of a high energy neutrino and an electron. These neutrinos can have energies up to 14.06 MeV and so can be detected in a chlorine detector.

Raymond Davies and his colleagues built a huge underground chlorine detector in the Homestake Gold Mine in South Dakota, USA in the late 1960s (shown in Figure 2.4(*b*)). It consisted of a tank containing 400 000 litres of the chemical perchloroethylene, C_2Cl_4, which is similar to cleaning fluid, and which contains a large number of chlorine atoms. The experiment ran for 20 years and the good news is that solar neutrinos from the decay of boron-8 nuclei were detected – the bad news is that only about a quarter of the expected numbers of neutrinos were measured (Figure 2.4(*a*)). This is the famous *solar neutrino problem* and it remains one of the great problems of modern astronomy.

Confirmation that the flux of high energy neutrinos indeed comes from the Sun has been provided by the Japanese Kamiokande II neutrino-scattering experiment. This large underground experiment, located in the Kamioka metal mine in the Japanese Alps, was originally designed to search for the products of the decay of protons, but was upgraded to act as a directional detector for neutrinos with energies greater than 7.5 MeV. When a high energy neutrino collides with an electron, the recoil direction of the electron is measured and this provides information about the arrival direction of the neutrino. The Japanese scientists found

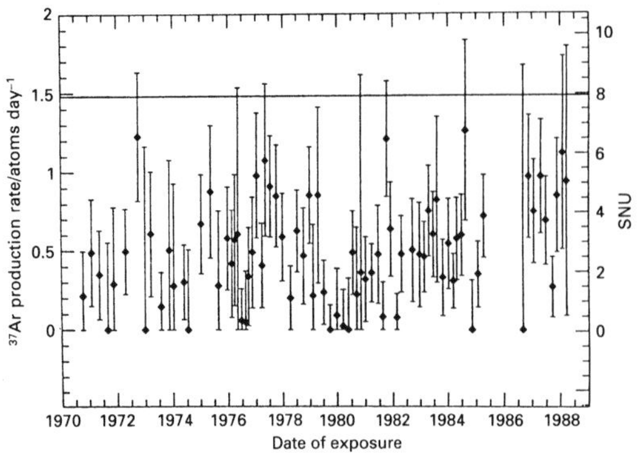

(*a*)

(*b*)

Figure 2.4. (*a*) A comparison of the observed flux of solar neutrinos from observations carried out at the Homestake Gold Mine through the period 1970 to 1988. The average production rate of radioactive argon, 0.462 ^{37}Ar atoms per day, is significantly smaller than the expected production rate of about 1.5 atoms per day. (*b*) The ^{37}Cl detector developed by R. Davis Jr and his colleagues. The photograph shows the tank which held 400 000 litres of perchloroethylene in a cavity 1500 m below ground at the Homestake Gold Mine.

a small but significant excess flux of neutrinos coming from the direction of the Sun (Figure 2.5). This is confirmation that the neutrinos originate in the Sun but their flux is again smaller than expected, being only about 46% of the predictions of the standard solar models.

A key test of the nuclear reactions which power the Sun, is the detection of the low energy antineutrinos associated with the first reaction of the p–p chain. This is a technically challenging experiment because the best substance to use as the detector material is the metal gallium and large amounts of it, in a highly purified state, are needed. When a solar antineutrino interacts with a gallium nucleus, a radioactive isotope of germanium is created, and so the number of germanium atoms produced in the gallium detector is a measure of the solar neutrino flux. Fortunately, the threshold energy for this interaction is only 0.23 MeV and so the low energy neutrinos from the Sun can be detected. Two experiments have been carried out using gallium as a detector. One of these is the GALLEX project (Gallium experiment) which

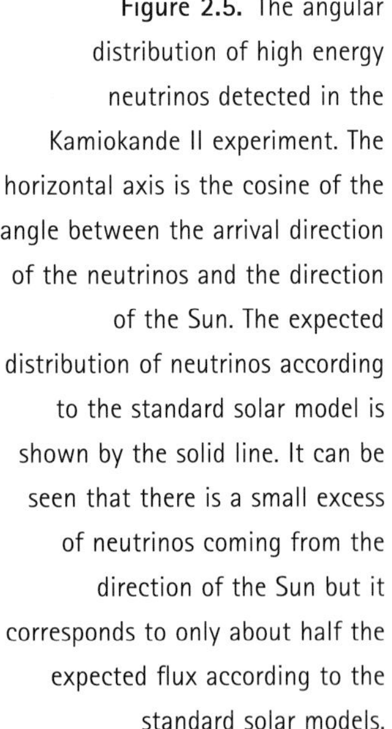

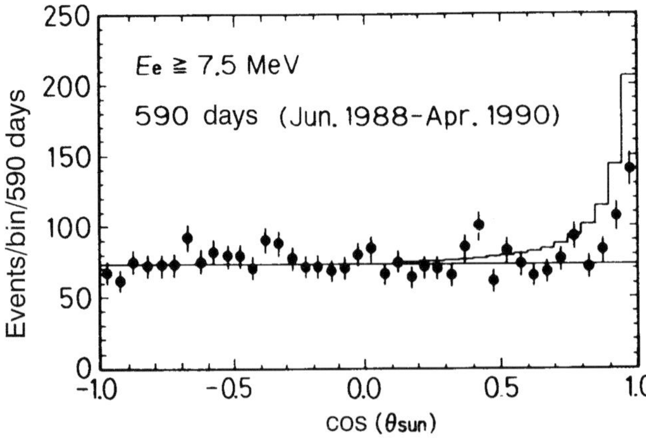

Figure 2.5. The angular distribution of high energy neutrinos detected in the Kamiokande II experiment. The horizontal axis is the cosine of the angle between the arrival direction of the neutrinos and the direction of the Sun. The expected distribution of neutrinos according to the standard solar model is shown by the solid line. It can be seen that there is a small excess of neutrinos coming from the direction of the Sun but it corresponds to only about half the expected flux according to the standard solar models.

is primarily a European collaboration with US and Israeli participation and which is located in the underground Gran Sasso Laboratory in Italy. The other is the SAGE project (Soviet American Gallium Experiment) which is primarily a project sponsored by the former Soviet Union with American participation and which is located in the Baksan Valley in the Caucasus Mountains. In June 1992, the first results of the GALLEX experiment were reported, the neutrino flux being 83 ± 19 (stat) ± 8 (syst) Solar Neutrinos Units (SNU), compared with an expected value of 134^{+20}_{-17} SNU. The quoted errors associated with the observed flux refer to statistical (stat) and systematic (syst) uncertainties. This is a crucial result because, although there is still a discrepancy, a neutrino flux of roughly the expected intensity has been detected from the essential first reaction of the p–p chain. In fact, this first reaction is responsible for most of the energy liberated in the p–p chain. A similar result was reported in 1994 from the SAGE experiment.

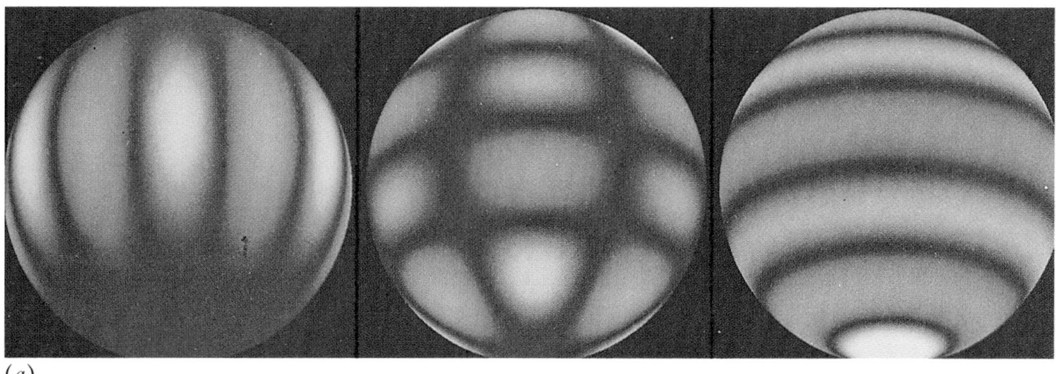
(*a*)

(*b*)

Figure 2.6. (*a*) A schematic diagram showing some of the simplest modes of sound wave oscillation of the Sun. In this representation, the green regions are moving outwards and the yellow regions inwards. (*b*) A schematic diagram of one of the higher modes of oscillation of the Sun, illustrating how these sound waves penetrate into the interior of the Sun. The blue regions are moving outwards and the red regions inwards. The different modes of oscillation probe the physical conditions to different depths in the Sun.

Thus, the detected fluxes of both high and low energy neutrinos are still low compared with the predictions of the standard model of the Sun, but the discrepancies may not be as great as they were once thought to be. It remains a challenge to understand the origin of these discrepancies.

The second remarkable development of the last 15 years has been a completely new way of studying the internal structure of the Sun and the stars. We are all familiar with the melodious sounds which are produced by musical instruments such as bells, gongs and xylophones. These sounds are made up of sound waves of different frequencies, which correspond to the natural frequencies of vibration of the instrument. If the frequencies of vibration and their relative intensities are measured, it is possible to work out a great deal about the physical structure of the instrument. For example, small bells produce high pitched sounds while large bells produce low pitched sounds and, from the analysis of these sounds, the size and structure of the bell can be estimated. In exactly the same way, if the Sun is perturbed, it resonates at its natural frequencies of vibration – some of the simplest modes of oscillation are shown in Figure 2.6(*a*). In fact, the Sun is constantly being perturbed in its interior by turbulent and convective motions originating in the outer part of the Sun, as illustrated in Figure 2.6(*b*), which shows how a particular high order mode of oscillation penetrates to some depth within the Sun.

The solar oscillations are very small indeed. Typically, the velocities associated with these vibrations amount to no more than about 0.4 km s^{-1} at maximum and extremely sophisticated instruments are required to detect these tiny motions on the surface of the Sun. These resonant modes of the Sun have now been measured and the new discipline of *solar*

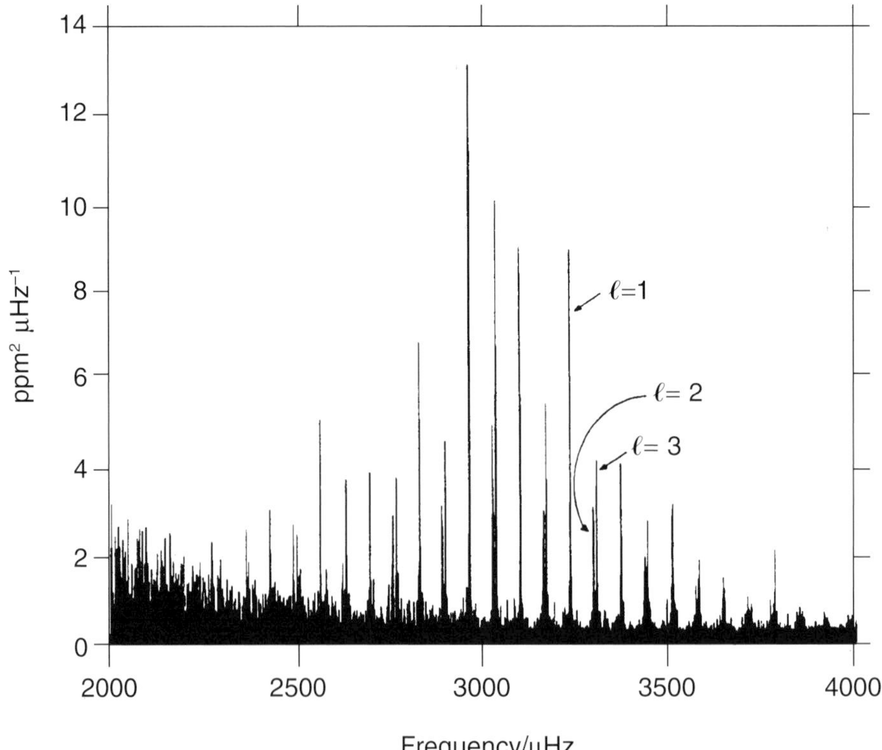

Figure 2.7. An example of the frequency spectrum of solar oscillations showing the fine structure splitting which contains a great deal of information about the internal structure of the Sun.

seismology, or, preferably *helioseismology*, has become one of most exciting fields of current astronomical research. An example of the results of one experiment to measure the resonant modes of the Sun is shown in Figure 2.7. This remarkable spectrum was obtained by T. Toutain and C. Fröhlich, who built a helioseismic detector for the Soviet Phobos spacecraft. The measurements were made during a period of 160 days, while the space vehicle travelled from the Earth to the vicinity of Mars. It can be seen that there are strong resonances in the vibrational spectrum of the Sun and a vast amount of detailed fine structure. By studying the details of these resonances of the Sun as a whole, its internal structure can be deduced.

An example of the precision with which the internal structure of the Sun can now be determined by this technique is shown in Figure 2.8. It turns out that the simplest property of the solar interior, which can be deduced most directly from the helioseismic observations, is the variation of the speed of sound with radius throughout the Sun. The speed of sound is related to the ratio of pressure and density at each point in the Sun and so is a convenient physical property which can be compared with the predictions of theoretical solar models. The speed of sound is also closely related to the square root of the temperature of the material of the Sun. This comparison is shown in Figure 2.8. It can be seen that the helioseismic data enable the speed of sound to be determined right into its very central regions. In Figure 2.8(*a*), the expected variation of the square of the speed of sound with radius is shown for a variety of different assumptions about the properties of the Sun, for example, the abundance of helium and its age. The model labelled $Y_0 = 0.28$, meaning that the helium abundance is 28% by mass, and age $t = 4.6 \times 10^9$ years agrees with the observations to within the thickness of the line. A more detailed comparison is shown in Figure 2.8(*b*), which shows the deviations of the inferred speed of sound throughout the Sun from the expectations of this standard solar model. It can be seen that theory and observation agree to better than 1% throughout the Sun. Theorists are now working intensively to improve the agreement between theory and observation. What is intriguing about these studies is that, not only are the internal properties of the Sun being determined in unprecedented detail, but new insights are also being gained into the behaviour of matter in bulk at the temperatures and densities found inside

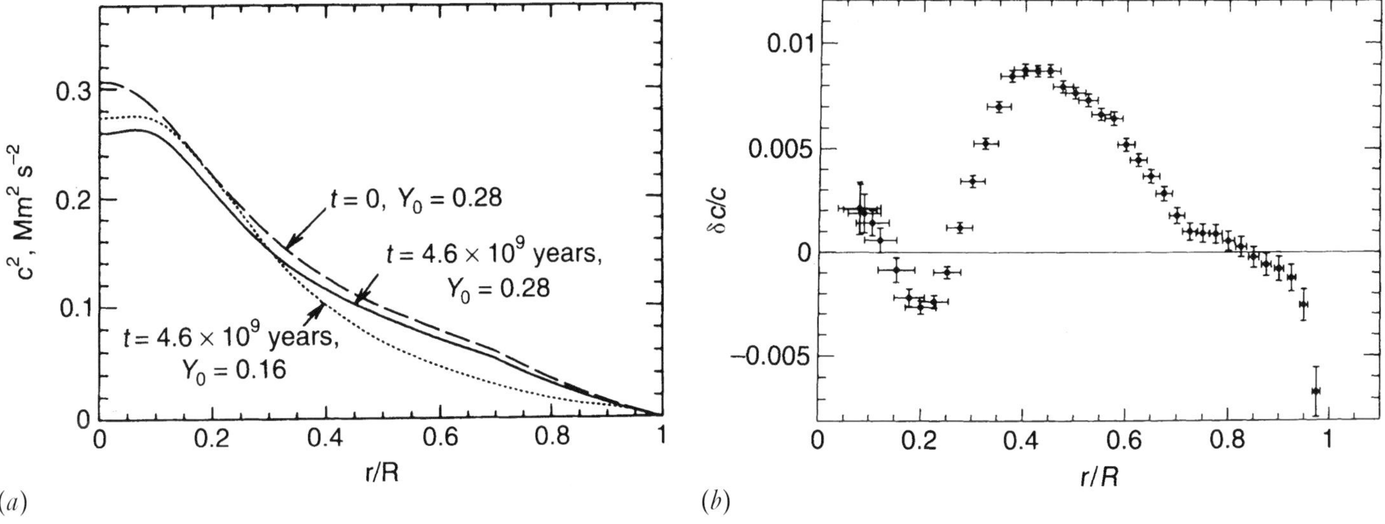

Figure 2.8. (*a*) Theoretical models for the variation of the square of the speed of sound inside the Sun with radial distance from its centre. The fractional abundance by mass of helium is denoted by the symbol Y_0 and the age of the solar model t in years is also shown. The observations agree with the line labelled $Y_0 = 0.28$ and $t = 4.6 \times 10^9$ years to within the thickness of the line. (*b*) Fractional deviations of the derived variation of the speed of sound $\partial c/c$ through the Sun as a function of radius for the best fitting model shown above. The points with error bars show values of the speed of sound derived by inverting the helioseismic data. It can be seen that these are in agreement with the standard model to better than 1% throughout the Sun.

the Sun. In the most recent studies, the spectrum of solar oscillations has been used to infer the internal rotation speed at different radii and latitudes within the Sun. These results are very important for understanding the properties of the outer convection regions of the Sun as well as the origin of the solar magnetic field and the 11-year solar cycle.

These techniques provide a new method for studying not only our own Sun, but also bright nearby stars. Astronomers are already attempting to observe the same types of oscillations in nearby stars using the largest ground-based telescopes. Space observatories are also being planned to provide the long periods of uninterrupted observation needed to detect the slow oscillations of bright stars under the perfect observing conditions of space. It must be emphasised that stellar seismology, or *asteroseismology*, on even nearby stars is very difficult and requires very large telescopes and a great deal of observing time. This is, however, the way to the future in the study of the internal structure and evolution of the stars.

The great achievement of helioseismology is that the astrophysics used to determine the internal structure of the Sun results in models which are in excellent agreement with observation. This is a very important result for understanding the origin of the solar neutrino problem, because it suggests that we can have confidence that the astronomers have provided the nuclear and particle physicists with the correct physical conditions within which the problem has to be solved.

2.3 The evolution of stars and the great cosmic cycle

Studies of the astrophysics of the Sun are of central importance for all astronomy. The Sun is by far the brightest and nearest star we can study and the theory of stellar structure has to account for its properties or else there is not much point in attempting to explain the properties of more distant stars. The theory of the Sun is sufficiently well understood for the same types of physics to be used to understand the structures and evolution of different types of star. Stars like the Sun are very stable, in the sense that they burn their nuclear fuel steadily in their cores and the energy released diffuses out through the stellar envelope, ending up as the light we observe from their surfaces. Most of the visible stars in the Universe are in a similar state of steady hydrogen burning. In stars less massive than about one and a half times the mass of the Sun, the p–p chain is the principal source of energy for the stars, whereas for more massive stars, the conversion of hydrogen into helium takes place through a different sequence of reactions known as the carbon–nitrogen–oxygen, or CNO, cycle. In this sequence of reactions, carbon acts as a catalyst for the formation of helium through the successive addition of protons to carbon nuclei, forming some of the rarer isotopes of carbon, nitrogen and oxygen in the process. This period of steady nuclear burning of hydrogen into helium is by far the longest phase of a star's active lifetime and stars like the Sun, for example, remain in this state for about 10 billion years.

To compare the theory of stellar structure and evolution with the observed properties of stars, one of the most powerful tools is the diagram first plotted by the Danish astronomer Ejnar Hertzsprung and the American astronomer Henry Norris Russell, in which the luminosities of stars are plotted against their surface temperatures. Appropriately, this diagram is known as the Hertzsprung–Russell, or H–R, diagram. In practice, it is more convenient to plot the *colour* of the star, rather than its surface temperature. From the considerations of Section 1.5, it can be appreciated that if we measure the ratio of intensities of a star in two different wavebands, for example, the red and blue regions of the spectrum, we obtain information about its surface temperature. Astronomers call the logarithm of this ratio the colour of the star. One of the most important recent space missions for all astronomy has been the HIPPARCOS project (High Precision Parallax Collecting Satellite) of the European Space Agency which has measured very precisely indeed the positions, distances and colours of bright stars (Figure 2.9(*c*)). For the first time, the H–R diagram for nearby stars has been determined with accurate distances, and hence accurate luminosities, and also with good statistics. The resulting H–R diagram is shown in Figure 2.9(*a*).

It can be seen that most of the stars in Figure 2.9(*a*) lie along a continuous sequence which runs from the bottom right to the top left of the diagram – this sequence is known as the *main sequence* and the stars belonging to it, known as *main sequence stars*, are all in the prolonged state of steady burning of hydrogen into helium, described above. In Figure 2.9(*b*), a 'theorist's' H–R diagram, in which luminosity is plotted against surface temperature, is shown and the various sequences present in Figure 2.9(*a*) are labelled. Along the main sequence, the masses of the stars are shown and it can be seen that they increase systematically along it from bottom right to top left. Our Sun is a very average star, lying about halfway along the main sequence.

There is another 'branch' of stars extending from the main sequence to the top right of the diagram. This sequence is known as the *giant branch*, since these stars are luminous but rather cool. In order to radiate so much energy, they must be of very great physical size. In old star clusters, there is also a branch which stretches across the H–R diagram at high luminosities which is known as the *horizontal branch*. Below the main sequence, there are a few very compact, intrinsically faint stars known as the *white dwarfs*, which we will discuss in Chapter 3.

The theory of stellar structure can give an excellent account of these features of the H–R diagram, which is one of the principal diagnostic tools for stellar evolution. The stars on the main sequence have masses between about one tenth and 60 times the mass of the Sun, a remarkably narrow range compared with their luminosities, which range from about 100 000 times the luminosity of the Sun for the most massive stars to only about one thousandth of the solar value for the lowest mass stars. There are good astrophysical reasons for this. The temperatures shown on the horizontal axis of Figure 2.9(*b*) refer to the surfaces of the stars but, just as in the case of the Sun, the temperatures in their centres are very much greater than these values. Models of the internal structures of the stars indicate that the surface temperatures are more or less proportional to their central temperatures. Now, the low mass stars towards the bottom right of the main sequence have much smaller surface temperatures than the Sun and correspondingly smaller central temperatures. Therefore, if the mass of the star is too small, the central temperature will be too low to sustain the nuclear reactions which convert hydrogen into helium. Theoretically, it is expected that stars with masses less than about one-fifteenth the mass of the Sun are too cool in their centres to burn hydrogen into helium. These low-mass objects are expected to be rather inert bodies, not unlike large planets, asteroids or huge lumps of rock. Collectively, they are known as *brown dwarfs*. One of the major unsolved problems in astronomy is how much of the mass in galaxies and in the Universe as a whole is tied up in objects like brown dwarfs. There could, in principle, be a great deal of matter in this form but it is very difficult to observe because the objects are cold, having no internal energy sources. They are expected to be very dim, infrared objects, the only source

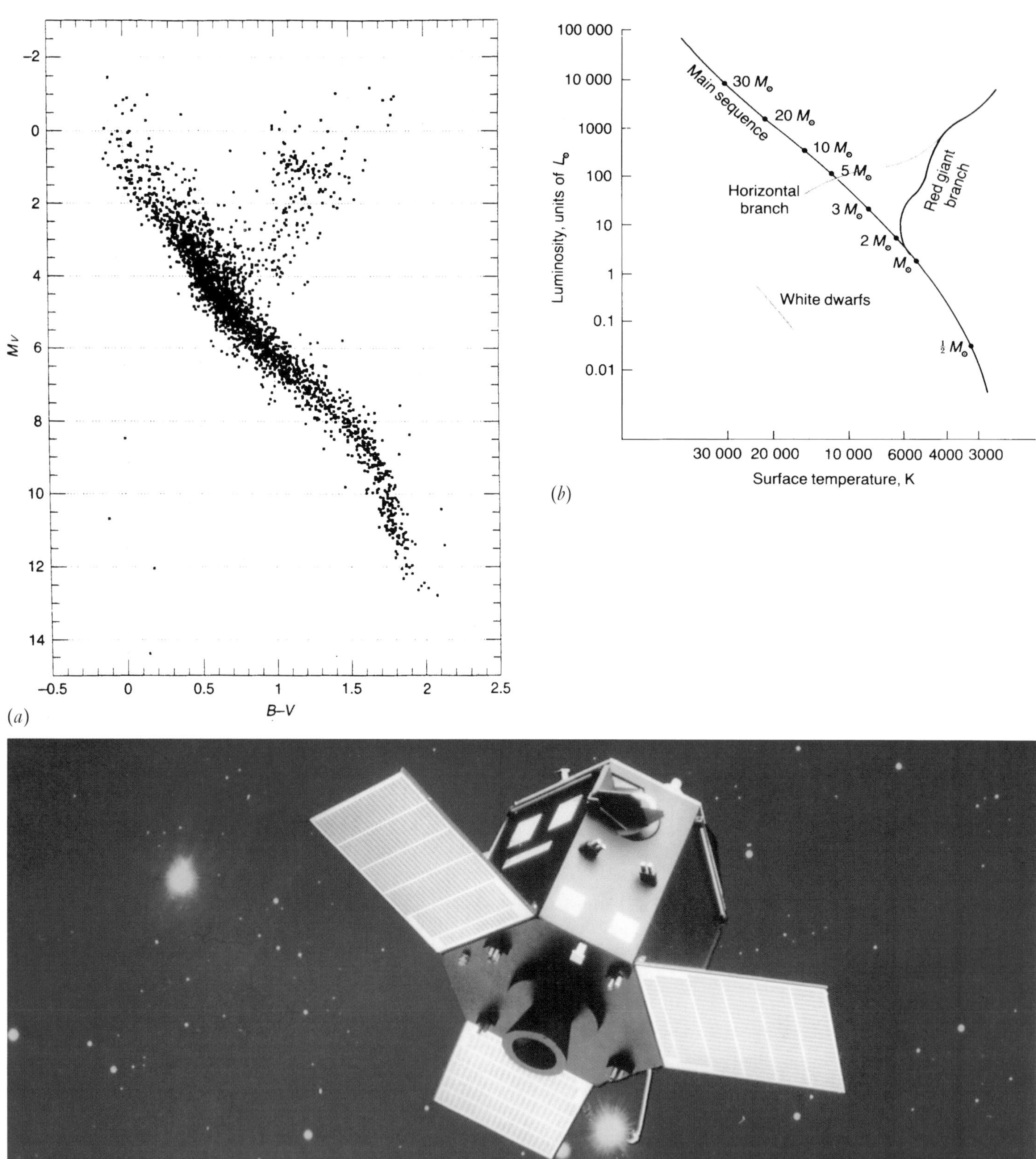

-2
0
2
4
6
8
10
12
14
M_V
-0.5
0
0.5
1
1.5
2
2.5
B–V
100 000
10 000
1000
100
10
1
0.1
0.01
Luminosity, units of $L_\odot$
Main sequence
30 $M_\odot$
20 $M_\odot$
10 $M_\odot$
5 $M_\odot$
3 $M_\odot$
2 $M_\odot$
$M_\odot$
$\frac{1}{2}$ $M_\odot$
Horizontal branch
Red giant branch
White dwarfs
30 000 20 000 10 000 6000 4000 3000
Surface temperature, K

(a)

(b)

(c)

Figure 2.9. (opposite) (*a*) The luminosity–colour diagram for nearby stars, also known as the Hertzsprung–Russell (H–R) diagram. The luminosities plotted up the vertical axis are absolute magnitudes measured in the standard *V* waveband, M_V. The colours are measured by the differences in apparent magnitudes in the standard *B* and *V* wavebands. This diagram has been derived from observations of 2927 stars in the solar neighbourhood by the HIPPARCOS satellite of the European Space Agency. All the stars plotted have distances determined to better than 10%. Stars do not occupy all regions of this diagram but form various sequences. (*b*) A schematic version of the H–R diagram showing the various sequences. Most stars lie along the band which stretches from the bottom right to top left of the diagram, which is known as the *main sequence*. The masses of stars at various positions along the main sequence are indicated. The *giant branch* extends from the main sequence to the top right of the diagram. The locations of the *horizontal branch*, observed in old star clusters, and the region in which white dwarf stars are found are also shown. (*c*) The HIPPARCOS astrometric satellite of the European Space Agency, which has measured precise positions for over 100 000 stars, as well as the distances used to construct Figure 2.9(*a*).

of energy being the internal thermal energy with which they were born. We will return to the problem of brown dwarfs in Chapter 4.

At the high mass end, stars with masses greater than about 50 to 100 times the mass of the Sun ($M_\odot$) are unstable. These are enormously luminous stars, about 100 000 times more luminous than our Sun, and, as a result, most of the internal pressure is provided by the pressure of radiation rather than by the pressure of hot gas. If this pressure becomes too great, the outer layers of the star are blown off and this instability limits the mass of main sequence stars to less than about 50 to 100 $M_\odot$. These very massive stars are so luminous that they burn up their nuclear fuel very rapidly and so their lifetimes are no more than about 1 to 10 million years. Thus, there are excellent physical reasons why all the stars we observe in the Universe should have masses in the range about 0.1 to 50 $M_\odot$.

During the long period it remains on the main sequence, a star continues to burn hydrogen into helium in its core but, when about 12% of its mass has been converted into helium by nuclear burning in its centre, the star becomes unstable. The central regions of the star contract while the outer layers expand. The central regions continue to contract and heat up, while the envelope expands to an enormous size creating a star which is many thousands of times larger in radius than the main sequence star from which it formed. The expansion of the star is only halted when its outer envelope becomes fully convective and then the star becomes a fully fledged *red giant*. In red giant stars with masses roughly the mass of the Sun, the nuclear burning of hydrogen into helium takes place in a shell about an inert helium core. In more massive stars, the central temperature of the star can become high enough for the nuclear burning of helium into carbon to take place. The precise sequence of events depends somewhat upon the mass of the star but the general picture is that the star moves off the main sequence and crosses the H–R diagram to the region occupied by the red giant stars.

One important aspect of this picture for the evolution of stars is that all the stages subsequent to the star's long period on the main sequence take place over a very much shorter time-scale than the long period it remains on the main sequence. The physical reason for this is that the luminosity of the star becomes very much greater when it evolves onto the giant branch and so the available nuclear fuel is burned up much more rapidly. Thus, if a cluster of stars was formed at some time in the past and the cluster is observed at a later time, say, 100 million years after it was formed, it is expected that all the stars with lifetimes less than this age will have evolved off the main sequence and become giant stars. Therefore, the main sequence should only extend up to that mass and luminosity of star with main sequence lifetime equal to the age of the cluster. Figure 2.10 shows the H–R diagrams for a number of star clusters of different ages and it can be seen that what is known as the *main sequence termination point* provides an estimate of the age of the cluster. A useful point of reference is that the Sun is predicted to have a main sequence lifetime of about 10 billion years. Since the age of the Universe is about this same age, it follows that in the oldest systems, the main sequence termination point will have evolved down to masses roughly equal to the mass of the Sun.

Among the most important clusters to be studied using these techniques are the globular clusters, an example of which, the star cluster 47 Tucanae, is shown in Figure 2.11(*a*). The globular clusters are among the very oldest stellar systems and form part of the old bulge

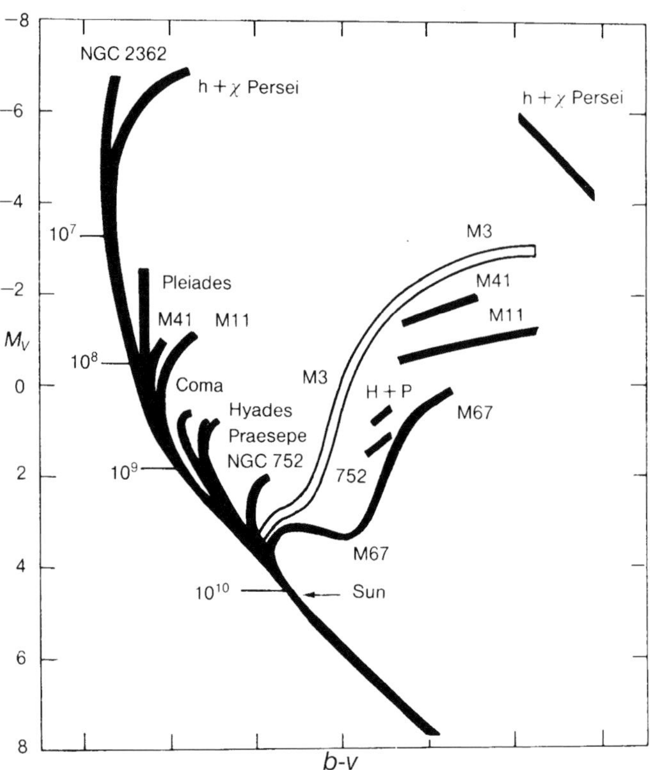

Figure 2.10. The H–R diagrams for star clusters of different ages. The differences in their H–R diagrams can be attributed to their different ages. The age corresponding to the main sequence termination points of stars of different masses are indicated in years along the main sequence. The youngest cluster is NGC 2362 and the oldest M67.

population of our Galaxy. Typically, they contain about one million stars. The H–R diagram for 47 Tucanae is shown in Figure 2.11(*b*). It can be seen that the H–R diagram for the stars of the cluster is very beautifully defined and a precise comparison can be made between the observed diagram and the predictions of theoretical models for the evolution of the stars in the cluster. This analysis has been carried out by John Hesser and his colleagues and some examples of their fits of the predicted distributions of stars to the data are shown in Figure 2.11(*b*). They find that the H–R diagram is consistent with the cluster having an abundance of elements such as carbon and oxygen which is only about 20% of their solar abundances and with ages which are about 12 to 14 billion years. The most recent studies suggest that the oldest stellar systems in the Galaxy may be even slightly older than this, about 16 billion years. This is a very important result for cosmology since it sets a lower limit to the age of the Universe.

In massive stars, the central temperatures are greater than that of the Sun and so the sequence of exhaustion of the nuclear fuel followed by contraction until a new nuclear energy source switches on can proceed through to the burning of carbon and oxygen to form heavier elements such as silicon. In the most massive stars, the process of nuclear burning can proceed through to the formation of iron, the most tightly bound of all the chemical elements. As each new phase of nuclear burning is switched on, the star undergoes considerable internal rearrangement of its structure. The star moves progressively up the giant branch shown in Figure 2.9(*b*) with brief excursions across the diagram as the different processes of nuclear burning are switched on.

In addition, as stars evolve up the giant branch, mass loss is believed to take place from

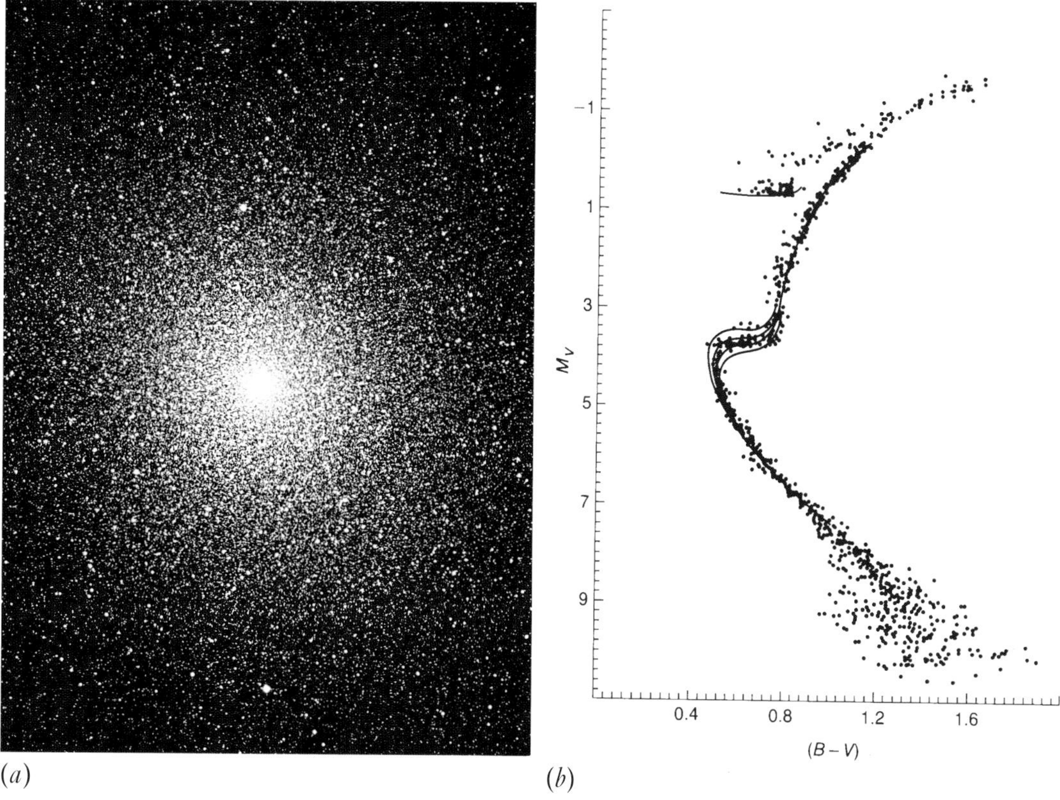

Figure 2.11. (*a*) The globular cluster 47 Tucanae. This cluster contains about one million stars and is a member of the bulge population of our Galaxy. The distribution of the stars in the cluster is spherically symmetric. (*b*) The H–R diagram for 47 Tucanae. The solid lines show different 'isochrones' for theoretical models for the distributions of stars on the H–R diagram. The term isochrone means that the theoretical distribution is what would be expected if the H–R diagram were observed at a given time after the formation of the cluster. The abundance of the heavy elements is only 20% of the solar value and the isochrones shown have ages of 10, 12, 14 and 16 billion years.

their surface layers, revealing their hotter central regions, and so the stars move to the left across the H–R diagram. In the case of solar mass stars, this mechanism is believed to be responsible for the origin of the horizontal branch stars which is shown schematically in Figure 2.9(*b*) and can also be seen in the H–R diagram of 47 Tucanae in Figure 2.11(*b*). These stars then evolve back towards the tip of the giant branch, a region occupied by long period variables and unstable stars. In the case of the massive stars, mass loss can be so strong that the stars are shifted a long way across the H–R diagram to the vicinity of the main sequence. It has been observed that there is an absence of stars of the very highest luminosities in the region of the red giants and this can be plausibly attributed to the effects of mass loss from the most massive giant stars.

Eventually, the star runs out of nuclear fuel and the central regions collapse. In the most massive stars, it is likely that the collapse is very violent and results in an enormous explosion in which the star ends up as some form of dead star – either as a *white dwarf*, a *neutron star* or a *black hole*. In less massive stars, such as the Sun, it is expected that the death throes are less dramatic. Once these stars reach the tip of the giant branch, they become unstable and blow off their outer layers to form the objects known as *planetary nebulae* while the core of the star shrinks, ultimately forming a white dwarf (Figure 2.12). We will discuss the physics of these different types of dead star in more detail in the next chapter. For our present purposes, what is important is that, when stars die, they eject a substantial fraction of their mass back into the space between the stars – what is known as the *interstellar medium*. Thus, in the course of stellar evolution, hydrogen is converted into heavier elements, the star

Figure 2.12. The planetary nebula known as the Helix Nebula. As the star dies, it blows off a shell of gas, while the core of the star collapses to become a compact helium star. The helium star is very hot and its intense ultraviolet radiation illuminates the ejected shell of gas. The helium star evolves rapidly to become a white dwarf.

evolves onto the giant branch and then, as it dies, the processed material is returned to the interstellar medium. The most massive stars, which can burn their nuclear fuel through to heavy elements such as carbon, silicon and iron, are responsible for enriching the interstellar medium with heavy elements when they explode. The next generation of stars is then formed from interstellar gas, which has been enriched by the products of stellar nucleosynthesis.

We will demonstrate in Section 2.4 that stars form in cool, dense regions of the interstellar gas. We can therefore picture the birth, life and death of stars as part of what I call the *great cosmic cycle*. Figure 2.13(*a*) shows this cosmic cycle from the birth of the star in a dense cloud of dust and gas through to its death as some form of dead star. This picture is like a time-lapse film, similar to those used to show plants growing and flowering in a matter of seconds – the trick is to open the shutter of the camera only once an hour or so. In Figure 2.13(*a*), the shutter of the 'camera' has been opened once every few million years. The star condenses rapidly out of an interstellar gas cloud and then settles down to a very long period of steady hydrogen burning as a main sequence star like our Sun. Our own Sun is about halfway through its life as a main sequence star, being about 4.6 billion years old, compared with a total lifetime of about 10 billion years. We therefore don't have to worry about running out of solar energy for some little time. It can also be seen that, as soon as the star moves off the main sequence to become a giant star, the final stages of evolution up the giant branch take place very rapidly compared with the star's main sequence lifetime. Since their lifetimes are so short, red giant stars are much less common per unit volume than main sequence stars but, because they are so luminous, they are not difficult to find and appear in considerable numbers in samples of bright stars. The evolutionary path of a solar mass star on the H–R diagram is shown schematically in Figure 2.13(*b*), the regions in which strong mass loss takes place being indicated by dashed lines.

(*a*)

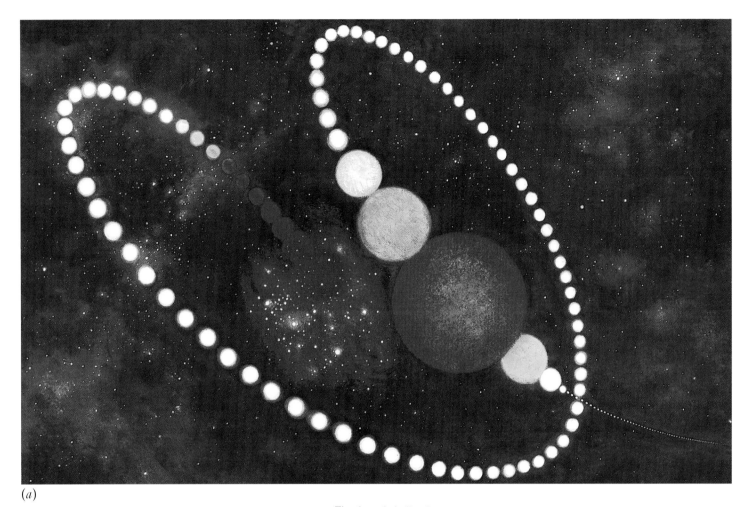

Figure 2.13. (*a*) A 'time-lapse' picture of the evolution of a star from its formation in a giant molecular cloud, through the main part of its lifetime as a main sequence star and ending in its expansion to form a red giant star and its subsequent demise as some form of dead star. The interval between images of the star corresponds to a few million years. (*b*) A schematic diagram illustrating the evolutionary path of a solar mass star on the H–R diagram. Those stages of evolution during which there is strong mass loss from the surface are indicated by dashed lines.

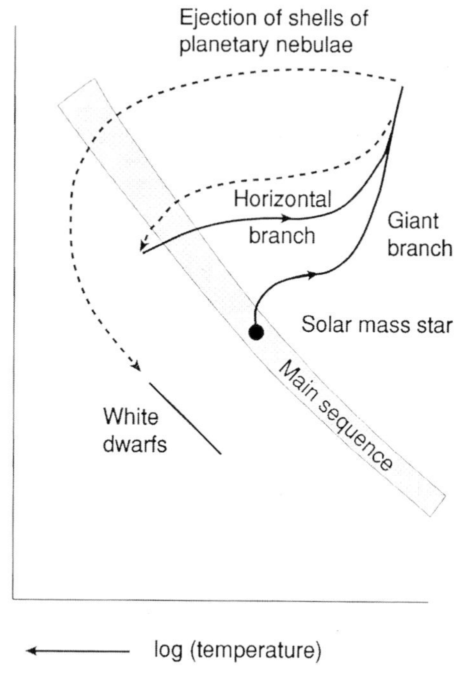

(*b*)

2.4 How to observe stars forming

One of the big problems of astronomy and cosmology, which we will tackle in Chapter 4, is to understand the origin of galaxies and the large scale structure of our Universe. We cannot, however, work out the sequence of events which took place when the galaxies first formed, and how they have evolved since then, unless we understand what it is that determines the rate at which stars form out of the pregalactic and interstellar gas. Once stars begin to form in a young galaxy, this influences the ability of the galaxy to make more stars in different ways. On the one hand, star formation reduces the amount of gas available to make more stars. On the other hand, once the first generation of massive stars has formed and evolved to the end of their lifetimes, they return processed gas, enriched with heavy elements, to the interstellar medium and this makes it easier for the gas to cool and so create the next generation of stars. Therefore, the problem of understanding what determines the rate of star formation is intimately tied up with the problems of understanding the origin and evolution of the galaxies as well. Disentangling exactly what happens when stars form and how the process depends upon physical conditions are among the greatest challenges of modern astronomy. We have learned a great deal about these problems over the last 10 to 15 years, thanks to our ability to study the very earliest stages of star formation in completely new ways, in particular, by infrared, millimetre and submillimetre astronomy.

There is one obvious problem which can be appreciated from Figure 2.13(*a*) and that is that the process of star formation takes place very rapidly compared with the lifetimes of main sequence stars and therefore we have to catch the objects just as they are forming. One of the best ways of finding the very youngest stars is to search in the vicinity of regions which are known to contain young stars. The nearest stellar nursery for massive stars to the Earth is

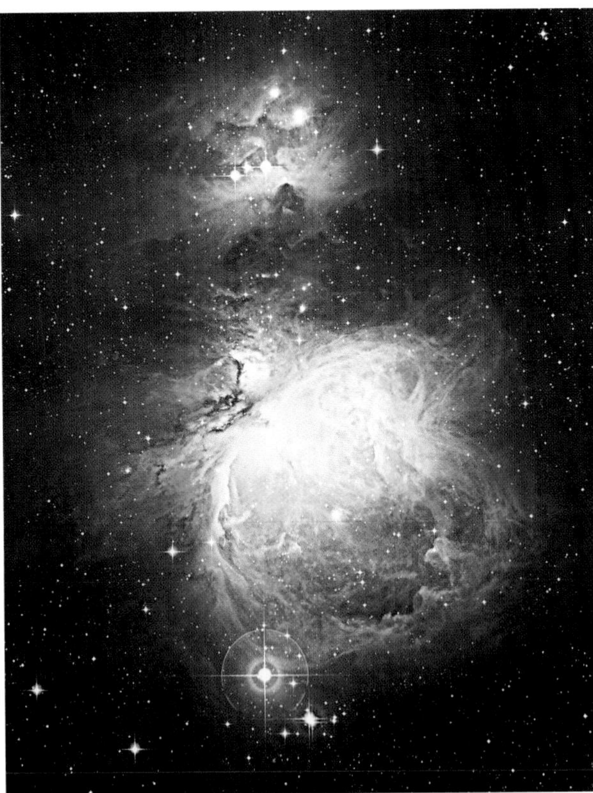

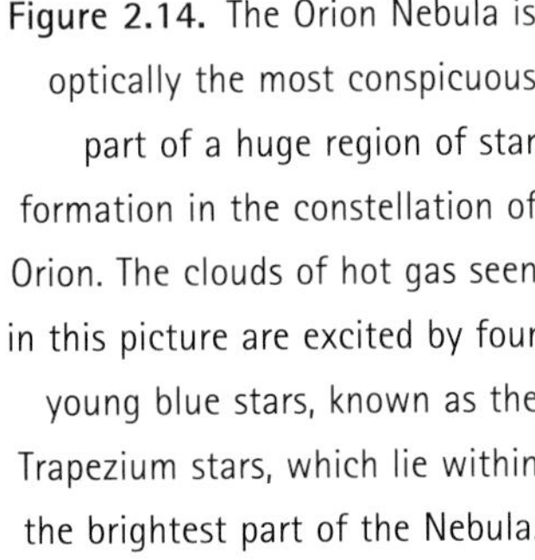

Figure 2.14. The Orion Nebula is optically the most conspicuous part of a huge region of star formation in the constellation of Orion. The clouds of hot gas seen in this picture are excited by four young blue stars, known as the Trapezium stars, which lie within the brightest part of the Nebula.

located in the constellation of Orion, which is one of the most famous constellations in the sky. Halfway down Orion's sword, there is a diffuse patch of light known as the Orion Nebula (see Figure 2.19). Observed with a large telescope, the Orion Nebula is one of the most beautiful objects in astronomy (Figure 2.14). From the point of view of star formation, the Orion Nebula is of special interest because we know that star formation has taken place in this region within the last million years or so. Some of these young massive stars are easy to identify. The four bright, blue stars in the brightest part of the nebula, which are known as the Trapezium stars, are no more than about a million years old and they are responsible for illuminating the beautiful filamentary structure which surrounds them. Although these stars are young in terms of the age of our Galaxy, they have already formed and what we are more interested in is observing stars at even earlier stages in their evolution. Ideally, we would like to study the stars throughout the period before the nuclear reactions were switched on in their centres.

There is one obvious problem with Figure 2.14 – whilst it is a very beautiful picture, it is very dusty. Dust is present everywhere in the vicinity of the Orion Nebula and prevents us observing inside the regions where the stars are forming. At first sight, it appears that dust is just a nuisance for astronomers but it turns out that it plays a crucial role in the formation of stars. We know that there is a considerable amount of dust in the plane of our Galaxy which prevents us observing to large distances in these directions (see Figure 1.7). Similar dust clouds are observed in spiral galaxies as well (see, for example, Figure 1.9). Often, these dust clouds obscure some of the most interesting regions we wish to study, for example, the centre of our own Galaxy. Physically, what is taking place is that the dust grains absorb or scatter the radiation incident upon them. From studies of the properties of dust grains, we know that the obscuration in the optical region of the spectrum is caused by small solid particles about 1 μm, that is, one millionth of a metre, in diameter, roughly the size of the particles of cigarette smoke. Carbon is one of the main constituents of the dust grains but silicon is also present in some grains.

Fortunately, there is a way of seeing through the dust and that is to make observations at longer wavelengths, that is, in the infrared region of the spectrum at wavelengths greater than 1 μm. In the infrared waveband, the wavelength of the radiation is larger than the size of the dust grains and the radiation is hardly absorbed or scattered at all. It is only when the grains are larger than the wavelength of the radiation that they are strongly absorbing. The best way of demonstrating how this change of wavelength influences the appearance of a region of star formation is to compare observations of the Orion Nebula in the optical and infrared regions of the spectrum.

Infrared astronomy has been very difficult until recently because the equivalent of photographic plates, with which to make pictures of the sky, has not been available. Since 1987, however, astronomers have had access to infrared electronic array detectors of very high sensitivity, and these have enabled pictures to be taken in the infrared waveband for the first time (Figure 2.16(*b*)).

Let us see what happens when the Orion Nebula is observed in the infrared waveband. Figure 2.15 is a short-exposure optical picture of the central regions of the Orion Nebula showing clearly the four Trapezium stars, which were discussed above. This optical photograph was taken at a wavelength of about 0.5 μm, roughly the middle of the optical waveband. Now, let us compare this image with one made at wavelengths which are about three

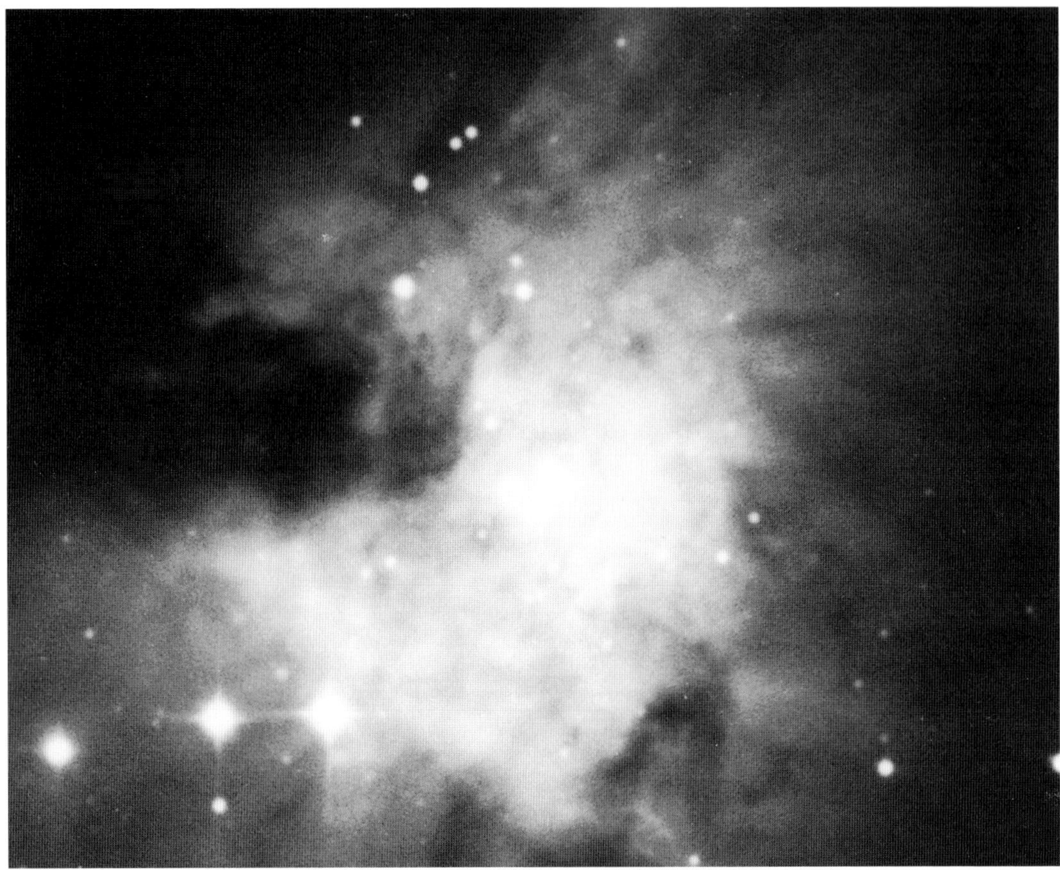

Figure 2.15. An optical image of the central regions of the Orion Nebula, showing the four bright Trapezium stars in the very centre of the nebula and the patchy obscuration by dust.

or four times longer than this optical wavelength. Figure 2.16(*a*) is a composite photograph of images made at 1.2, 1.65 and 2.2 μm. The effect is as if a veil has been removed from in front of the region. The patchy dust obscuration has disappeared and it is now evident that there is indeed a cluster of young stars present, roughly centred on the four bright central stars. The four bright, blue Trapezium stars are still there but there are also objects which have no counterpart in the optical image and these are of the greatest interest.

Figure 2.16(*a*) has been colour coded so that yellow and red objects are only observable in the infrared regions of the spectrum. The reddest regions are those which are the strongest infrared emitters. The most interesting features are the yellow and red regions to the north of the Trapezium stars where some enormously luminous infrared sources are found. The two brightest objects are known as the Becklin–Neugebauer object (to the north) and the Kleinmann–Low Nebula (to the south of BN), which were discovered in the 1960s by these pioneers of infrared astronomy. The remarkable feature of these sources is that they are very dusty indeed but, despite that, they are enormously powerful sources of far infrared radiation which must be produced from sources deeply embedded in these dust clouds. The sources to the north of the Trapezium stars have luminosities more than 100 000 times the luminosity of the Sun but that radiation is all emitted as infrared radiation at wavelengths longer than 2 μm, rather than in the optical waveband.

Why is the radiation so intense in the far infrared waveband? The answer is very simple.

(*a*)

(*b*)

Figure 2.16. (*a*) A composite infrared image of the Orion Nebula, taken by the Anglo-Australian Telescope. (*b*) The UK Infrared Telescope (UKIRT), which was the world's first thin mirror 4-metre telescope. The infrared camera can be seen located at the Cassegrain focus behind the primary mirror.

There are very young stars, possibly even stars in the process of formation, what are known as *protostars*, buried deep inside the dense dust clouds present in the Orion Nebula. These objects are sources of intense optical and ultraviolet radiation but there is so much dust surrounding them that this radiation is absorbed by the dust grains. In the process of absorbing the radiation, the dust grains are heated up and this heat is then reradiated by the grains at the temperature to which the dust has been heated. The temperature to which the dust grains are heated close to these sources turns out to be typically about 30 to 300 K. We can now use the relationship between temperature and wavelength (Figure 1.20) to work out the wavelengths at which most of the radiation is emitted. You will already have guessed the correct answer – the radiation is emitted at wavelengths between about 10 and 100 μm, corresponding to the infrared and far infrared regions of the spectrum, at which the dust particles are transparent to radiation. We can think of the dust as acting as a transformer, which enables the young star or protostar to get rid of its intense optical and ultraviolet radiation by reradiating these enormous luminosities at far infrared wavelengths, at which the radiation can escape unimpeded. This process is illustrated schematically in Figure 2.17.

These considerations put a completely new complexion upon the beautiful far infrared map of the Galaxy (Figure 1.28). Regions of star formation are among the principal contributors to that far infrared picture of our Galaxy and the physical process is precisely that described above, namely, the absorption and reradiation of optical and ultraviolet radiation by interstellar dust. We now know, from observations made by the IRAS satellite, that this process takes place in all types of galaxy in which star formation is observed. In particular, some of the most luminous galaxies in the Universe seem to emit huge amounts of radiation in the far infrared region of the spectrum. These are the objects known as *starburst galaxies*, in which enormous bursts of star formation seem to be taking place.

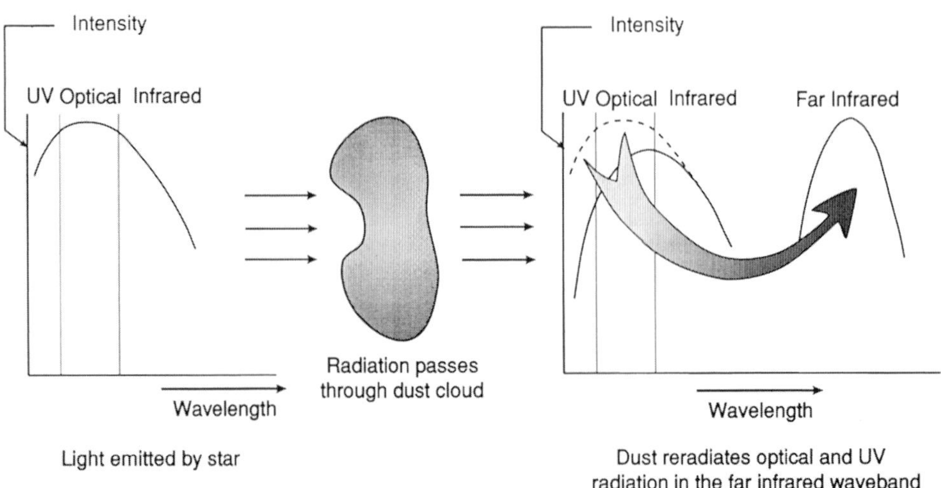

Figure 2.17. Illustrating the process of absorption of optical and ultraviolet radiation by dust grains and its reradiation in the far infrared region of the spectrum at the temperature to which the grains are heated.

2.5 The problems of star formation

The new observing capabilities described in the last section have revolutionised the study of the physical processes taking place in regions of star formation, but how are we to interpret these observations in terms of the theory of star formation? The basic problem is to understand how very diffuse gas, such as is found in the space between the stars, can be condensed into stars with typical densities about 10^{24} times greater than the interstellar gas. The answer is that the interstellar gas is unstable to large scale disturbances. There are all sorts of influences acting upon the gas which do not allow it to remain in a quiescent state. The most important of these is the effect of gravity and this leads us to the study of the behaviour of diffuse gas clouds under gravity.

Let us consider first an isolated gas cloud. If the cloud is massive enough, it will collapse under its own gravity. More precisely, if the force of gravity exceeds the repulsive forces associated with the internal pressure and turbulent motions within the cloud, collapse will take place. What is particularly important is that the collapse develops *exponentially*. What this means is that the cloud halves its radius in equal successive time intervals under the force of gravity so that we encounter a runaway situation in which the density of the cloud can grow very rapidly to very large values indeed in a finite time. This famous instability was discovered by James Jeans in 1902 and is appropriately known as the *Jeans instability*.

There is a simple way of illustrating exponential collapse under gravity. Suppose we take a long pointer with a sharp end and stand it on its tip. If we set it up very carefully, it will balance for a moment but it will fall over, even if it isn't pushed. Suppose the angle from the vertical direction is measured at equal time intervals. We find that the angle increases by the same proportion in equal time intervals as illustrated in Figure 2.18. This is what is called *exponential growth* and it has the important property that, even if we begin with infinitesimally small perturbations, the pointer (or gas cloud) still collapses under gravity to a large angle (or to a high density) in a finite time.

Therefore, if the gas clouds in the Galaxy are big enough, they are expected to collapse under gravity. There are various ways in which the collapse can get started and it is not clear which of these is the most important. For example, we know that the spiral arms in spiral galaxies contain large quantities of dust and gas and their gravitational influence can strongly

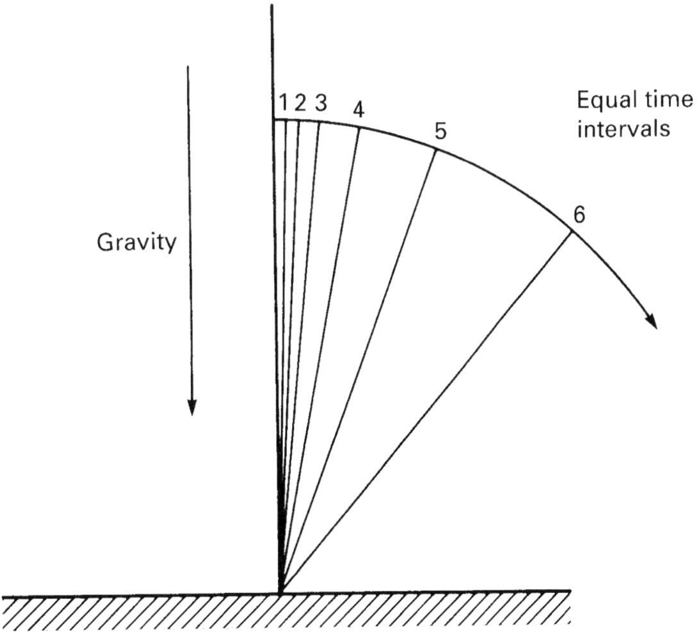

Figure 2.18. Illustrating the process of exponential collapse of a falling pointer under gravity. The angle from the vertical increases by the same fraction in equal time intervals.

perturb the gas distribution. The spiral arms can clump the interstellar gas to quite high densities and then the Jeans instability can take over. Dense gas clouds can also be created if a diffuse gas cloud is hit by the expanding shell of an exploding star. Consequently, there are several ways in which the density of an interstellar gas cloud can be significantly enhanced over its average density.

Are these dense gas clouds observed? The answer is 'Yes'! In typical interstellar gas clouds, dust is expected to be present as well as gas and the combination of these is very effective in cooling the clouds to low temperatures. The mechanism by which diffuse gas clouds cool is exactly the same as that by which protostars and young stars embedded in dust clouds lose energy, that is, by the radiation of heated dust. The clouds cool to such low temperatures, about 10 to 30 K, that the gas is expected to be in molecular rather than atomic form. We can therefore search for the characteristic line radiation of these molecules, which is emitted in the millimetre and submillimetre wavebands, as can be inferred from Figure 1.20.

The discovery of *giant molecular clouds* in our Galaxy has been one of the most important advances of modern astronomy. Figure 2.19 is a photograph of the constellation of Orion, superimposed upon which are contour lines showing the intensity of the millimetre line radiation of the interstellar molecule carbon monoxide, CO. It can be seen that, if our eyes were sensitive to millimetre radiation rather than light, there would be a huge patch of intense emission on the sky larger than the constellation of Orion in this direction. We now know that there is as much gas in the disc of our Galaxy in the form of molecular hydrogen as there is in the form of atomic hydrogen, what is referred to as neutral hydrogen. The molecules are mostly confined to giant molecular clouds, such as that present in the constellation of Orion. The typical density of the giant molecular clouds is about 100 to 1000 times that of the diffuse interstellar gas. Dust plays a crucial role in shielding the molecules from inter-

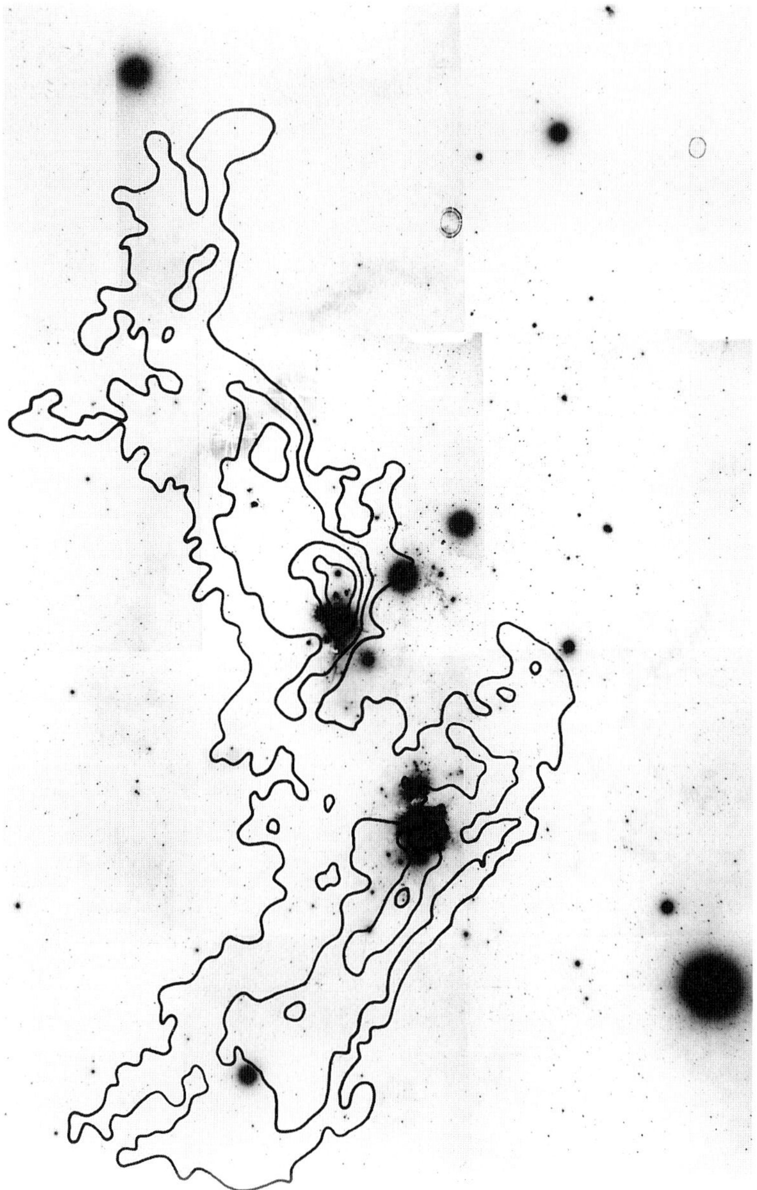

Figure 2.19. A composite photograph of the constellation of Orion showing Orion's belt and sword. The intense region of emission half way down Orion's sword is the Orion Nebula. Superimposed on this photograph are contour lines showing the intensity of millimetre line emission due to the molecule carbon monoxide, CO. The giant molecular clouds are of enormous size and have mass about a million times the mass of the Sun. Intense star formation occurs in the densest regions of the molecular clouds.

stellar optical and ultraviolet radiation, which would otherwise dissociate these fragile objects into their constituent atoms. It is also important that there is a great deal of fine structure in the giant molecular clouds and that the most active regions of star formation occur in the densest regions of the clouds, as can be observed in Figure 2.19.

We can now refine the questions to be answered. How is it that, once the interstellar gas is clumped into cool giant molecular clouds, stars are formed? This issue is far from resolved, but there are a number of important clues about what must happen. There are three big problems to be solved. The first is that, in order to make stars, the gas cloud has to lose the energy it gains as it collapses. As a gas cloud collapses, the kinetic energy of collapse heats up the interior of the cloud until the pressure of the hot gas might increase to such an extent as to prevent it collapsing any further. This is known as the *energy problem*. Therefore, there must exist effective mechanisms by which the heat energy is radiated away, as the star collapses. Dust plays a vital role in this process, because it is quite extraordinarily effective in getting rid of the heat generated in the interior of the protostar. As the interior of the protostar heats up, it radiates optical and ultraviolet radiation which is absorbed by the dust grains and transformed into far infrared radiation which can escape from the protostar. This is exactly the process illustrated in Figure 2.17. Thus, although dust prevents us observing optically inside these dense regions, it plays a pivotal role in enabling stars to condense to very high densities from diffuse interstellar matter. This is almost certainly the solution to the energy problem and accounts for the observation that regions of star formation are such powerful sources of far infrared radiation.

Although we may have solved the energy problem, we have to consider the dynamics of the collapsing cloud as well. The second big problem arises from the fact that there is a great deal of turbulence in the regions where stars are formed and therefore any piece of the cloud which starts to collapse under gravity almost certainly acquires some rotation. As the rotating cloud collapses, the rotation is amplified. This is exactly the process by which an ice skater is able to spin so rapidly. To begin with, the skater spins slowly with arms outstretched and then, when the arms are drawn in, the rate of rotation increases enormously. Another way of demonstrating this important effect is to hold a weight in each hand and sit on a rotating stool with arms outstretched. Then, ask a friend to make the chair rotate and, when you pull your arms in, your rotation rate will increase.

Exactly the same process occurs when stars collapse. Any rotation is vastly amplified because of what is known as the *law of conservation of angular momentum*. This is the law which causes the skater or the person sitting on a rotating chair to spin up. The problem for collapsing stars is that, if the rotation is amplified too much, it can halt the collapse of the star. This is exactly the force you feel when standing on a rotating roundabout. If you try standing upright when the roundabout is rotating, you feel a force pushing you outwards – this is an apparent force due to the rotation of the roundabout and is called the centrifugal force. In stars, the rotation can become so great during collapse that the centrifugal force can halt the collapse of the protostar and prevent the star forming. There must therefore exist efficient ways of getting rid of the rotation or, more precisely, the angular momentum, of the collapsing material. This problem is known as the *angular momentum problem*.

There is a third problem, in some ways similar to the angular momentum problem, associated with the fact that there is very likely to be a weak magnetic field present in the collapsing cloud. From a variety of observations, we know that there is a weak magnetic field in interstellar space. Its field strength amounts to only about one hundred thousandth of the Earth's magnetic field but it turns out to be dynamically important in understanding the properties of the interstellar medium. Similar weak magnetic fields have been measured in the regions in which stars are forming.

It was Michael Faraday who demonstrated that the behaviour of magnetic fields can be understood in terms of the properties of magnetic lines of force. Figure 2.20 shows the familiar picture of the magnetic lines of force about a magnetic dipole. According to Faraday's description, the strength of the magnetic field is proportional to the number of field lines passing through unit area and so the field is strongest at the poles of the magnet and weakest at the equator.

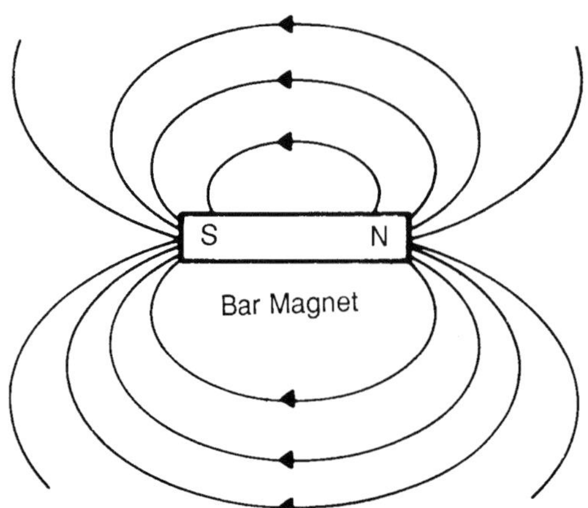

Figure 2.20. Illustrating Faraday's concept of magnetic field lines about a magnetic dipole. The field strength is proportional to the number of field lines per unit area perpendicular to the lines of force.

Magnetic fields play an important role in star formation because the matter of the cloud is tied to the magnetic field lines. One of the important results of plasma physics is that, if there are sufficient free electrons present in a gas which contains a magnetic field, the electrons are strongly tied to the magnetic field lines. One way of envisaging this process is to note that, if a charged particle, such as a free electron, moves through a magnetic field, the magnetic (or Lorentz) force acting on the particle causes it to move in a spiral path about the magnetic field direction. Therefore, if the magnetic field lines move, the electrons have to move with them. The same is true of the ions as well. It turns out that, even if there is a very small number of free electrons in an interstellar gas cloud, this is sufficient to tie the magnetic field to the gas in the cloud. We know that there are free electrons in the clouds, probably produced by collisions between high energy particles and the atoms and molecules in the clouds. Furthermore, the neutral gas is tied to the electrons and ions in the cloud by collisions between the neutral and charged particles. Thus, however the gas moves, the magnetic field lines are dragged with it. This idea has been formalised in a very beautiful theorem, which tells us that we can think of the magnetic field lines as being frozen into the gas, a phenomenon known as *magnetic flux freezing*.

The problem associated with the magnetic fields of collapsing stars is illustrated schematically in Figure 2.21. The magnetic field lines are frozen into the rotating disc of gas. If the disc were to rotate as a solid body, the pattern of the magnetic field lines shown in Figure 2.21(*a*) would be preserved. In order to ensure solid body rotation, however, the speed of rotation of the disc has to increase exactly in proportion to distance from the centre. In fact, it is more likely that the outer annuli do not rotate as rapidly as the inner annuli but rather lag behind the velocity distribution required for solid body rotation. Figure 2.21(*b*) shows what is expected to occur if, instead, the speed of rotation of the disc is the same at all radii. It can be seen that, as the disc rotates, the field lines are stretched in the direction of rotation. In addition, since the strength of the magnetic field is proportional to the number of field lines per unit area, as the magnetic field distribution is wound up, the number of field lines per unit area increases and so the magnetic field strength increases inexorably because of the rotation of the gas cloud. This is a real problem because it means that eventually so much magnetic energy could build up that the collapse would be halted. This is known as the *magnetic field problem*.

There is no agreed solution to the last two problems, namely, the means by which the protostar gets rid of its angular momentum and its magnetic fields, but there is one remarkable discov-

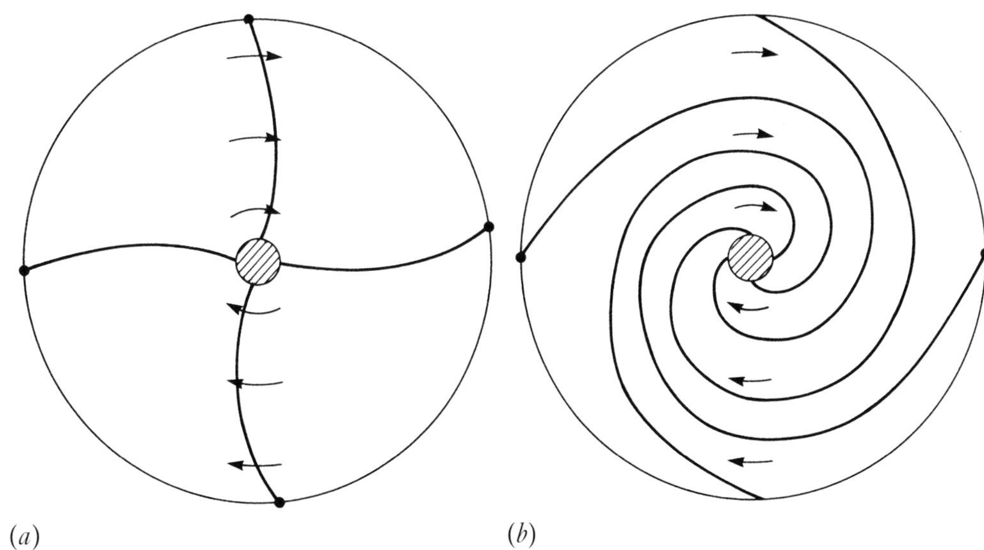

Figure 2.21. Illustrating the amplification of magnetic fields because of magnetic flux freezing. (*a*) The initial configuration of magnetic field lines in a rotating disc. (*b*) The magnetic field configuration if the magnetic field is frozen into the rotating disc and the speed of rotation is constant at all radii. The effect is to wind up the magnetic field lines and so increase the magnetic field strength.

ery which may well be important in understanding how they can be solved. It came as a complete surprise when jets of molecular material were discovered emerging from regions containing very young stars or protostars. Figure 2.22 shows millimetre images of two of these jets, which are known as *bipolar outflows*. The big surprise was that the molecular material is not collapsing onto the central object but is expelled in opposite directions at high speeds from a compact region in which a star is forming. The Doppler shifts of the molecular line emission have shown convincingly that one of the jets is coming towards us whilst the other is going away. A schematic picture of what is observed in these bipolar outflows is shown in Figure 2.23.

Bipolar outflows are found wherever young stars are being formed. It is not clear exactly what the structure of these flows is. In some cases, the jets are long and narrow, that is, they are said to be highly collimated. In others, the bipolar outflows are much less collimated and have an hour-glass appearance. There may be some relation to the lobes observed in the remarkable image of the very luminous young star η-Carinae, observed by the Hubble Space Telescope (Figure 2.24).

The origin of these bipolar molecular outflows is uncertain but it is tempting to associate them with the winding up of the magnetic field lines, as the rotating protostar collapses. The rotation of the cloud results in the formation of 'magnetic funnels' along the rotation axis of the protostar and these provide a natural axis along which material may be ejected. If the magnetic fields become too strong during the winding up process, the magnetic field lines may reconnect to relieve the strong magnetic stresses. When this process occurs, magnetic energy is dissipated, resulting in strong heating of the reconnection region. This type of process may well be important in creating bipolar outflows.

The rotation of the protostar not only creates a preferred axis along which the material can be ejected, but it also defines a plane perpendicular to the rotation axis, in which a rotating disc of dust and gas can form. Rotation can prevent collapse of the cloud perpendicular to the axis of rotation, but there is no reason why the infalling gas cannot form a disc by collapsing parallel to the axis of rotation. Evidence for such discs of molecular material has been observed about the axis of the bipolar outflow in the case of L1551 (Figure 2.22(*a*)). The thermal dust radiation from such discs has also been detected in the submillimetre waveband. Convincing

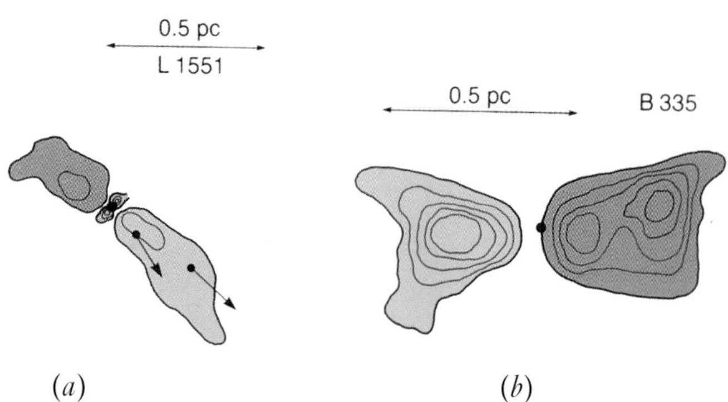

Figure 2.22. Two examples of high velocity bipolar outflows associated with very young stars. The young stars or protostars are indicated by black dots. The red lobe is moving away from the observer whilst the blue one approaches the observer. (*a*) The bipolar outflow in the source L1551. In this case there is also evidence for a molecular disc of denser gas perpendicular to the direction of the outflow. (*b*) The bipolar outflow source B335.

evidence for the presence of dusty discs about stars has been found by the IRAS satellite. In the case of the bright star Vega, the dusty disc extends to distances which are several times the size of our Solar System. A remakable image of this disc at an infrared wavelength of 3.5 μm has been obtained by masking the intense emission from the star itself (Figure 2.25). Evidence for similar dusty discs has been found about a number of stars similar to Vega.

One of the great surprises of the Hubble Space Telescope programme has been direct evidence for dusty discs about low mass stars in the vicinity of the Orion Nebula. Robert O'Dell noticed that, in the first images taken with the repaired telescope, there seemed to be haloes about some of the stars in the Nebula (Figure 2.26). He has interpreted these haloes as dusty discs which are illuminated by the luminous Trapezium stars in the centre of the Nebula. This discovery could only have been made by the Hubble Space Telescope because the angular sizes of the discs are only about 1 arcsec and so would have been difficult to resolve using ground-based telescopes.

The most exciting interpretation of these observations is that these dusty discs may represent the earliest stages of the formation of planetary systems about young stars. Once a disk of dusty material has formed, the solid particles come together to form larger and larger bodies and eventually planetary sized objects are formed by this process of coalescence. One of the great prospects for the next generation of large ground-based telescopes is the detailed study of protoplanetary discs about nearby stars. What many of us are hoping is that it will be possible to observe directly the process of planet formation by studying large numbers of circumstellar discs at different stages in their evolution about young stars.

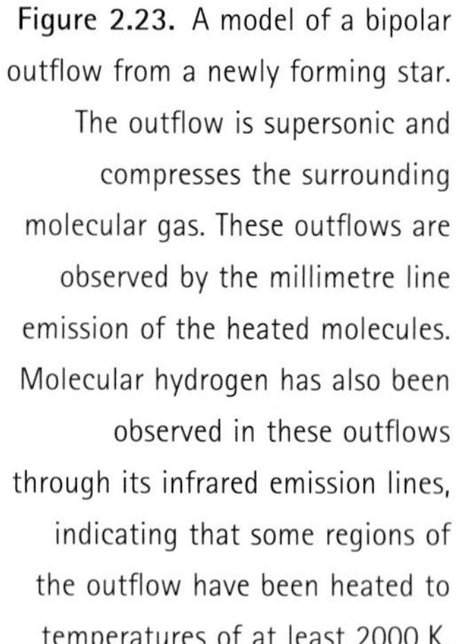

Figure 2.23. A model of a bipolar outflow from a newly forming star. The outflow is supersonic and compresses the surrounding molecular gas. These outflows are observed by the millimetre line emission of the heated molecules. Molecular hydrogen has also been observed in these outflows through its infrared emission lines, indicating that some regions of the outflow have been heated to temperatures of at least 2000 K.

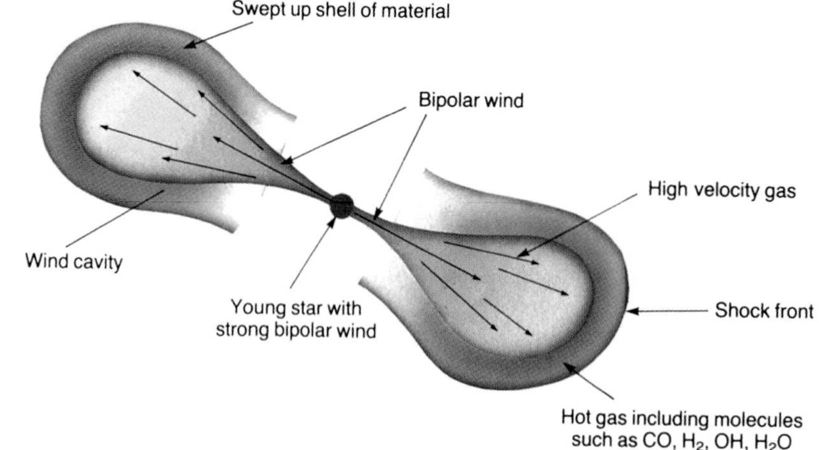

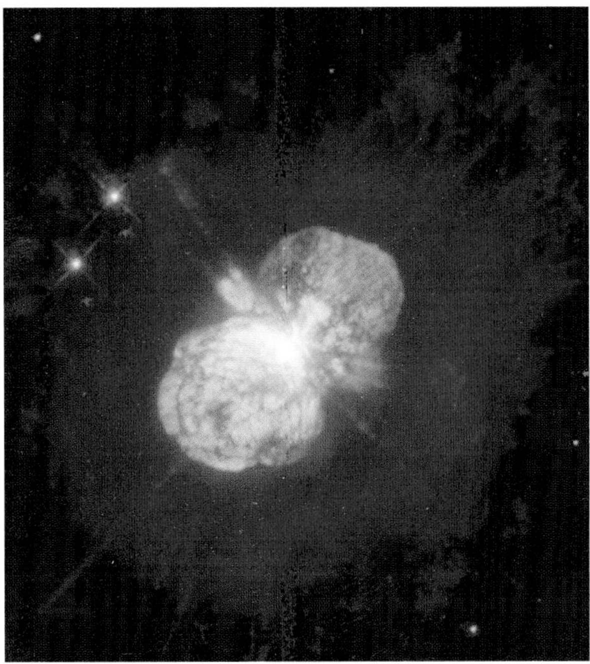

Figure 2.24. The massive young star η-Carinae as observed by the Hubble Space Telescope. This young star seems to have inflated large lobes, reminiscent of the bipolar outflows observed about young stars.

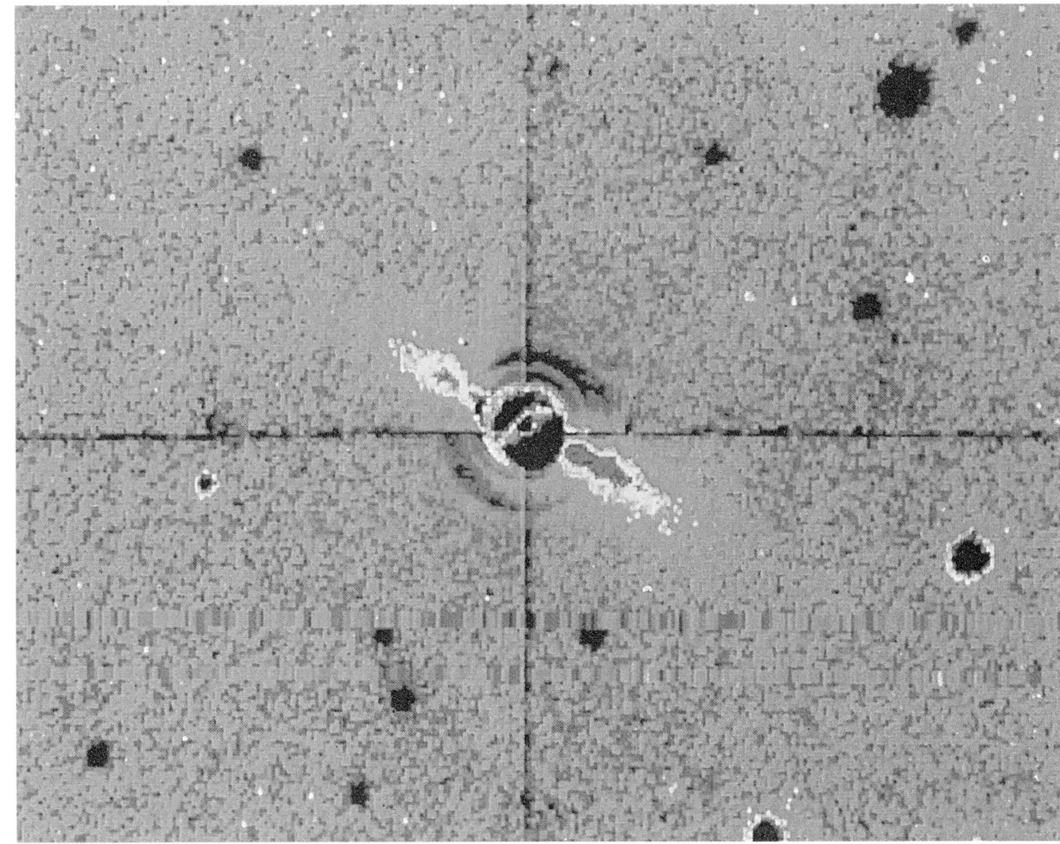

Figure 2.25. An image of the dusty disc about the bright star Vega (or β-Pictoris) at a wavelength of 3.5 μm obtained by Bradford Smith and Richard Terrile at the Las Companas Observatory in Chile. The star is so bright that a special mask has been used to prevent the radiation from the disc being swamped by light of the star.

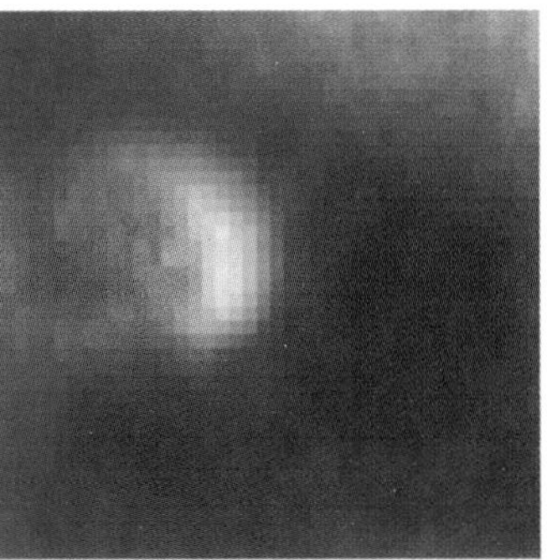
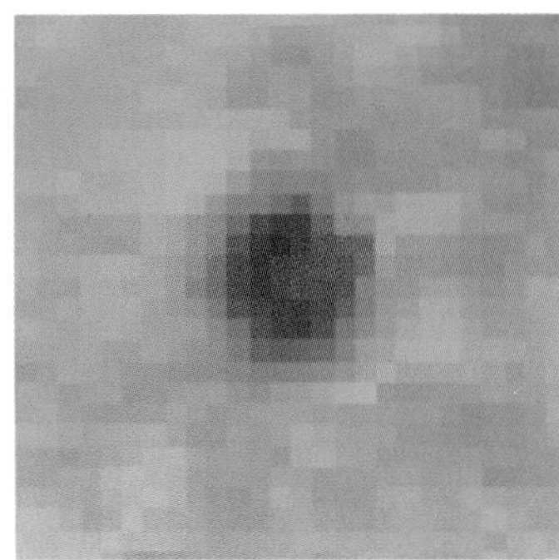

Figure 2.26. Optical images of dusty discs about low mass stars in the Orion Nebula. The discs are observed in reflected light from the luminous Trapezium stars at the centre of the Nebula.

There is one other important aspect of the discs which form about protostars and that is that they may provide a means of solving the angular momentum problem. When a disc forms by collapse of the rotating protostellar material, it must be in a state of rotation, because of the conservation of angular momentum. Once the disc has formed, it provides an effective means for removing angular momentum from the protostar. If the material of the protostar is linked to the disc by magnetic fields, angular momentum can be transported outwards by the stresses in the magnetic field and so the material can collapse onto the protostar. There are similarities between this process as applied to the discs of protostars and the physics of accretion discs, which are important in understanding binary X-ray sources and active galactic nuclei – these will be described in Chapter 3.

These ideas have been assembled into a standard picture of how stars form by Frank Shu and his colleagues (Figure 2.27). The picture begins with density inhomogeneities collapsing under their own self-gravity within a giant molecular cloud (Figure 2.27(*a*)). It can be shown that, in a spherically-symmetric collapsing cloud, the central regions collapse most rapidly to a high density and form a central core. Then the rest of the material of the collapsing cloud rains down onto the core of the protostar and so builds up the mass of the protostar. This phase is known as the accretion phase of the star formation and, during it, the source of energy is the kinetic energy of collapse of the material onto the compact core (Figure 2.27(*b*)). The energy liberated by the accreted matter is removed by the process of reradiation of the optical and ultraviolet radiation by dust in the far infrared waveband. Also during this phase, a rotating disc of material begins to form in a plane perpendicular to the rotation axis of the protostar. As the process of accretion continues, at some point, stellar winds burst out along the rotation axis of the accreting star, producing the characteristic bipolar outflows observed about protostars and young stars (Figure 2.27(*c*)). As the accretion continues, the mass of the protostar and its central temperature increase. Eventually, the central temperature of the protostar becomes high enough to sustain the nuclear burning of hydrogen into helium in the core of the protostar and the star begins its life as a main sequence star. The accretion of material onto the star ceases and what is left is a young hydrogen burning star and a rotating disc of cool dust and gas, out of which planetary systems can form (Figure 2.27(*d*)).

One of the more striking discoveries of the IRAS mission has been that objects with spectral characteristics corresponding to each of these phases of star formation have now been observed. It should be emphasised that there are many unknowns in this picture and the details of many of the physical processes involved are poorly understood. For example, we

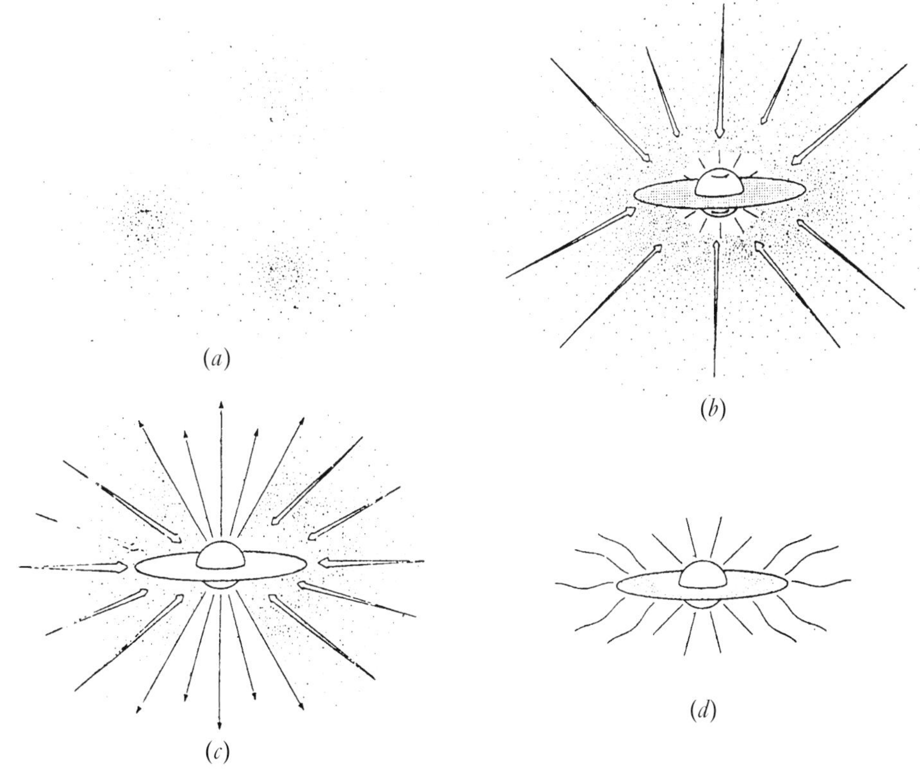

Figure 2.27. A schematic diagram illustrating a possible sequence of events which leads to the formation of stars with planetary discs, according to the picture developed by Frank Shu and his colleagues. The various phases of the star formation process are described in the text.

now know that a large fraction of the stars in the Galaxy are binary systems rather than single stars. Therefore, the theory of star formation has to take into account the fact that, during collapse, the formation of binary stars is favoured. Nonetheless, the picture of Shu and his colleagues is a useful aide-mémoire for the types of process which must occur during the process of star formation.

One of the more remarkable observations made by the Hubble Space Telescope has illustrated beautifully stage (*c*) of the above scenario, namely, the ejection of jets of material perpendicular to the accretion disc about a protostar. Chris Burrows and his colleagues have imaged the protostellar object HH30, in which the accretion disc about the embryonic star is observed edge-on (Figure 2.28). Since the disc is observed precisely edge-on, it obscures the young star itself. The light from the star, however, illuminates the top and bottom surfaces of the disc, thereby making them clearly visible. A reddish jet is ejected from the inner regions of the disc, quite possibly from the prototstar itself. Observations of the jet over a period of a year have shown that the structure moves away from the protostar at a speed of about 250 km s^{-1}.

2.6 Interstellar chemistry and the origin of life

One very intriguing aspect of the discs about young stars is the presence of molecules and these are needed in order to begin the process of formation of biological systems. One of the great discoveries of millimetre astronomy has been that the dense molecular clouds contain an enormous diversity of different types of molecule. Over 100 different molecular species have now been identified in the interstellar gas, ranging from simple molecules like

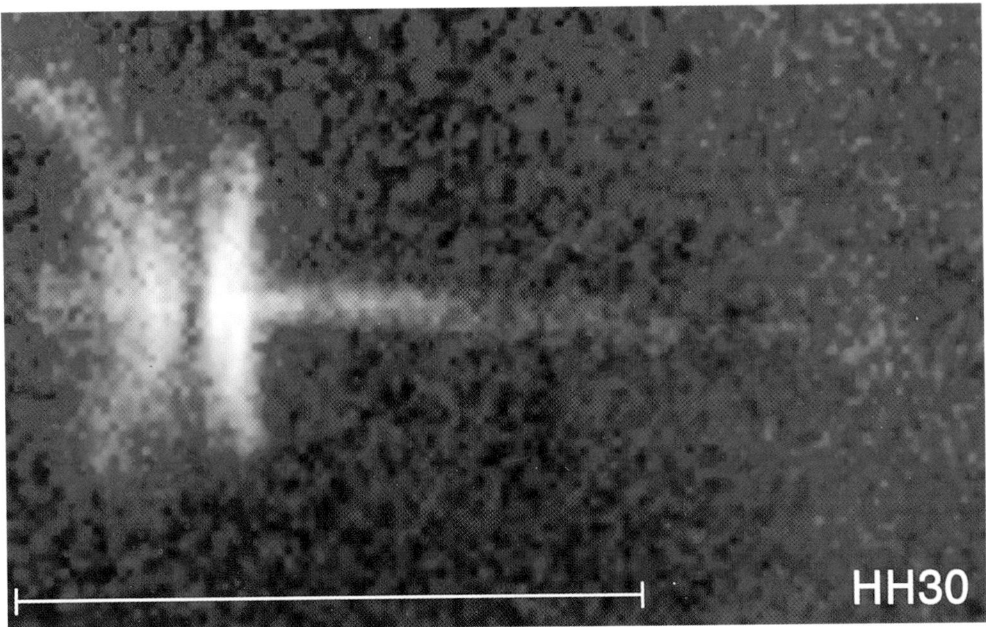

Figure 2.28. An image of the protostellar object HH30 as observed by the Hubble Space Telescope by Chris Burrows and his colleagues. The accretion disc is observed edge-on, and so the protostar itself is obscured. The reddish jet is ejected from the protostar at a speed of 250 km s^{-1}. The scale shown on the picture is a distance of 1000 times the distance between the Earth and the Sun, that is, 1000 astronomical units.

molecular hydrogen, H_2, carbon monoxide, CO, and the hydroxyl radical, OH, to complex molecules such as ethanol, C_2H_5OH, long acetylenic chains, ring molecules such as H_3^+, SiC_2 and C_3H_2 and so on. The existence of these molecules has given rise to the new discipline of *interstellar chemistry*. The conditions of interstellar space are very different from those found in chemical laboratories. Some molecular species have been found in interstellar space which are difficult to observe in the laboratory, because they are very reactive and hence have very short lifetimes under laboratory conditions. They can, however, survive in the very low density conditions of interstellar space. The result is that a number of ions and radicals which have not been observed in the laboratory have been observed in interstellar space through their molecular line emission.

Since many molecules can now be observed in molecular clouds, the understanding of molecular reactions in the rarified conditions of interstellar space has become an important part of the study of the interstellar medium. These studies enable the physical conditions in regions of star formation and in giant molecular clouds to be determined in completely new ways. Of particular interest are the processes by which the huge diversity of molecules found in interstellar space are formed. In some regions, gas phase reactions are thought to be important but many of the more complex molecules are believed to have been synthesised in reactions taking place on the surfaces of dust grains.

The complete list of known molecules includes everything needed to begin the process of biological synthesis. There is already present in the interstellar medium a rich molecular soup in which the biochemical molecules necessary for the creation of primitive life forms can be synthesised. In 1994 the simplest amino acid, glycine, was reported to be present in interstellar space. It is an open question whether or not these observations are relevant to the understanding of the origin of biological life on Earth. At least, if the biologists need them, these molecules are certainly present in substantial quantities in interstellar space.

3 The origin of quasars

3.1 The origins of high energy astrophysics

The most powerful energy sources we know of in the Universe are located in the centres of certain galaxies and these are referred to collectively as *active galactic nuclei*. It might seem somewhat perverse to discuss active galactic nuclei before we have discussed ordinary galaxies but there are good reasons for this. Many of the most important clues concerning the physics of active galaxies and quasars come from studies of the objects formed as end points of stellar evolution, which were mentioned briefly in the last chapter.

To set the scene, we recall that the principal components of normal galaxies are stars and gas. Most of the light of galaxies is starlight and this defines the overall appearance of a galaxy. In elliptical galaxies, a spheroidal distribution of stars is observed and the galaxy is held together by the mutual gravitational attraction of the stars for one another. In spiral galaxies, there is a central bulge, not unlike an elliptical galaxy in appearance, and a stellar disc, which is in a state of rotation. The centrifugal forces acting on the stars and gas due to the rotation of the disc are balanced by the gravitational attraction of all the matter in the galaxy. The discs of the spiral galaxies can contain a significant amount of gas, but, in the elliptical galaxies, there is generally very little gas at all. Many of these basic features of galaxies were established by Hubble in the 1920s and 1930s, soon after he had established the extragalactic nature of the galaxies in 1925.

Until about 1945, that seemed to be the end of the story – the galaxies are simply systems composed of large numbers of stars and gas clouds. After the Second World War, however, it gradually became apparent that there is much more to the story than this. In the early 1930s, Karl Jansky was employed at the Bell Telephone Laboratories at Holmdel, New Jersey, USA and was assigned the task of identifying naturally occurring sources of radio noise which were interfering with radio communications. In what turned out to be a classic series of observations made at the long radio wavelength of 14.6 m (20.5 MHz), he announced in May 1933 the discovery of the radio emission of our Galaxy. The observations were confirmed by the radio engineer and amateur astronomer Grote Reber who published a radio map of the Galactic plane in 1941. The nature of the radio emission remained, however, a mystery.

As a result of the development of radar during the Second World War, radiowave technology made enormous strides and, after the War was over, a number of the physicists, who had pioneered the development of radar, turned their attention to the origin of the Galactic radio emission discovered by Jansky. It was soon discovered that, in addition to the emission of our own Galaxy, which is beautifully illustrated in Figure 1.31(*a*), there also existed 'point' sources of radio emission away from the Galactic plane. Some of these sources of radio emission turned out to be associated with massive galaxies, many of them among the most massive galaxies known. The radio spectra of these radio sources were similar to that of our own Galaxy – unlike the spectra of stars, the radio spectra were continuous and featureless and bore no resemblance to the black body spectrum of hot bodies at all.

The nature of the Galactic radio emission was finally understood in the 1950s. The radiation is due to the process known as *synchrotron radiation*, and is caused by very high energy electrons spiralling in the magnetic field in the interstellar medium. As we described in Section 2.5, it is a general property of charged particles moving in a magnetic field that they spiral about the magnetic field direction. There is another fundamental law of physics which tells us that, when charged particles are accelerated, they radiate electromagnetic radiation. In a magnetic field, the electrons are continually accelerated towards the centres of their spiral orbits and so they emit radiation. Now, it turns out that the electrons responsible for the radio emission from the Galaxy are of very high energy indeed – their kinetic energy of motion, E, far exceeds their rest mass energy, $E_0 = m_e c^2$. These particles are called *ultrarelativistic electrons* and, because they are travelling so close to the speed of light, the radiation they emit is strongly beamed in the direction of motion of the electrons. When we analyse the radiation from such electrons, the spectrum is found to be smooth and continuous (Figure 3.1(*a*)). This type of radiation is observed in the powerful particle accelerators called *synchrotrons* and this is the origin of the term synchrotron radiation. The intensity of radiation depends upon the number of ultrarelativistic electrons and the strength of the Galactic magnetic field.

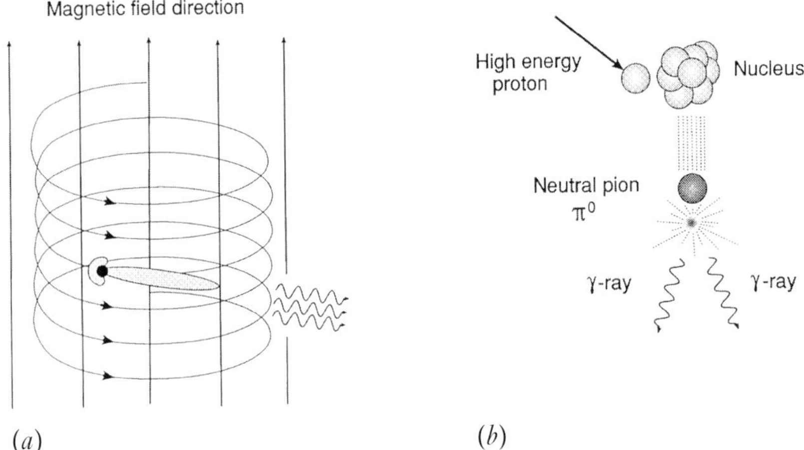

Figure 3.1. Illustrating the processes by which high energy particles can be detected astronomically. (*a*) Ultrarelativistic electrons emit highly beamed electromagnetic radiation when they spiral in a magnetic field. This form of radiation is known as synchrotron radiation and is responsible for the radio emission of our Galaxy. (*b*) When high energy protons or nuclei collide with the nuclei of atoms and molecules of the interstellar gas, the unstable particles known as pions are produced and the neutral pions, π^0, decay very rapidly into high energy γ-rays. These γ-rays have been detected by γ-ray telescopes such as the NASA Compton Gamma-Ray Observatory.

We can be certain that this process is responsible for the Galactic radio emission. The strength of the interstellar magnetic field has been determined and satellite experiments, conducted well above the Earth's atmosphere, have detected a flux of very high energy electrons, which originated in the local interstellar medium. When these two observations are put together, it turns out that the observed intensity of the Galactic radio emission can be explained. Thus, radio observations of galaxies enable us to determine where the high energy electrons and magnetic fields are located. The observation of strong radio signals from galaxies demonstrated convincingly that they must contain two other important ingredients in addition to stars and gas, namely, *very high energy particles* and *magnetic fields*. It is apparent from Figure 1.31(*a*) that the high energy electrons and magnetic fields must be present throughout the disc of our Galaxy. Synchrotron radiation has proved to be one of the most important diagnostic tools for detecting high energy electrons and magnetic fields, wherever they are encountered in the Universe and it dominates a great deal of thinking in what is now called *high energy astrophysics*.

These observations provided important clues concerning the origin of the particles known as *cosmic rays*. In 1913, Victor Hess discovered that the top of the Earth's atmosphere is constantly being bombarded by a flux of very high energy particles which, since they had to originate from above the Earth's atmosphere, were called cosmic rays. In the 1930s, Walter Baade and Fritz Zwicky made the inspired guess that these particles are accelerated in supernova explosions but there was no convincing evidence for this hypothesis at that time. The cosmic rays are mostly high energy protons and nuclei and the details of their energy spectra and chemical composition were only defined once cosmic ray telescopes were flown in satellites in the 1960s. The problem, from the point of view of the relation of these observations to astronomy, was that the cosmic rays only provide information about high energy particles in the vicinity of the Earth. Radio and γ-ray observations have, however, provided complementary information about the distribution of high energy particles throughout the Galaxy. The radio map of the synchrotron radiation of high energy electrons in the interstellar medium (Figure 1.31(*a*)) shows that there is a flux of high energy electrons present throughout the Galactic plane, similar to that observed at the top of the atmosphere. In the same way, the γ-ray map of our Galaxy (Figure 1.27(*a*)) shows us where the high energy protons and nuclei are located. When high energy protons and nuclei collide with the nuclei of atoms and molecules of the interstellar gas, the particles known as pions, π, are created. These are unstable particles and the neutral pions, π^0, decay almost immediately into high energy γ-rays (Figure 3.1(*b*)). Thus, the γ-ray emission of the Galaxy traces out where the cold gas and high energy protons and nuclei are located in the Galaxy. The similarity of the radio and γ-ray maps shows that the plane of the Galaxy is filled with a flux of very high energy protons, nuclei and electrons.

3.2 The radio galaxies and the discovery of quasars

The galaxies which were associated with sources of intense radio emission were named *radio galaxies* and the big surprise was that some of them had quite enormous radio luminosities. In 1954, the second brightest radio source in the northern sky, the source known as Cygnus A, was found to be associated with a very distant, massive galaxy and its radio luminosity was more than 100 million times that of our own Galaxy. This meant that there had to be enormous fluxes of high energy particles, and possibly stronger magnetic fields, in these radio galaxies as compared with those present in our own Galaxy. What is the origin of this vast flux of relativistic particles and magnetic field energy?

These results were remarkable enough but what made matters even more intriguing was the fact that the radio emission did not originate within the galaxy itself but from two enormous radio lobes located on either side of the galaxy (Figure 3.2). It appears as if the galaxy has ejected two huge clouds of radio emitting material from its nucleus. Since this phenomenon was discovered, there have been many superb studies of these objects using advanced radio telescopes, such as the Very Large Array (VLA), shown in Figure 3.3(*b*), located in the New Mexico Desert in the USA. It is now clear that the reason for these huge lobes of radio emission is that they are continuously supplied with high energy particles and magnetic fields by jets originating in the nucleus of the galaxy. The beautiful radio picture of Cygnus A (Figure 3.3(*a*)), made using the VLA, displays this process very clearly. The consequence is that there must be something very remarkable going on in the nuclei of radio galaxies – they are able to generate enormous amounts of energy, which is then expelled in the form of jets to power the huge radio lobes which extend far beyond the confines of the galaxy. Nothing

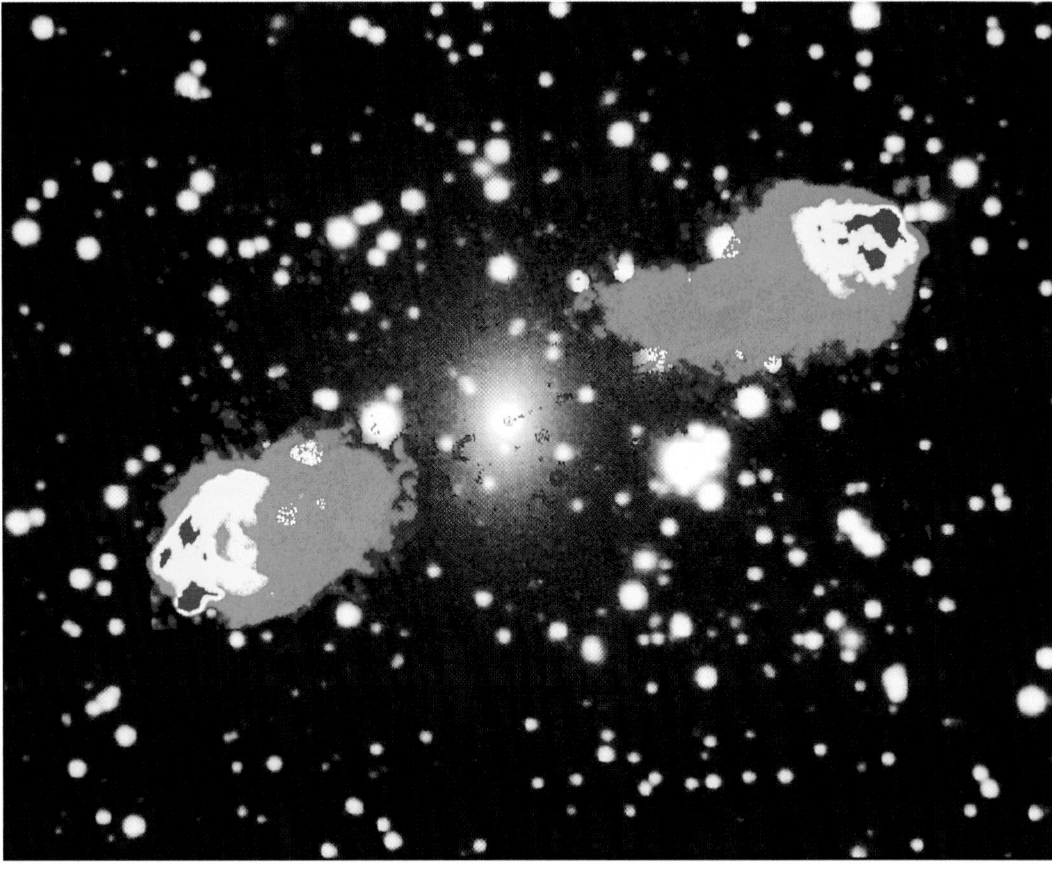

Figure 3.2. The radio structure of the bright radio source Cygnus A, shown in colour, superimposed upon an optical picture of the same area of sky. A massive galaxy at the centre of a rich cluster of galaxies lies between the huge radio lobes and its nucleus is the energy source for the extended radio structure.

like this had been expected theoretically, nor had anything like it been observed anywhere else in the Universe at that time.

This story was just beginning to unfold in the early 1960s, when an even more extraordinary discovery was made. One of the most exciting occupations of that time, in which I was involved as a research student at the beginning of my PhD programme, was the discovery of the distant galaxies associated with powerful radio sources. Many of the associated galaxies

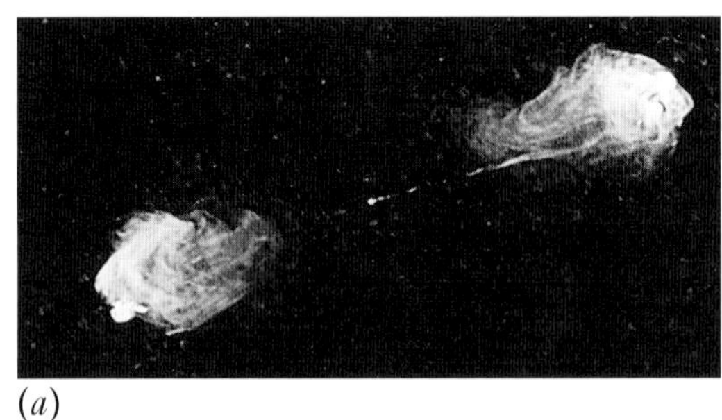

(*a*)

(*b*)

Figure 3.3. (*a*) The detailed radio structure of the radio source Cygnus A, as observed by the VLA. In addition to the radio lobes with their remarkable fine structure, there are intense 'hot-spots' towards the leading edges of the lobes, in which the energy density of radiation is very high. There is also a compact radio source in the nucleus of the radio galaxy and radio jets which channel the energy from the nucleus to the hot-spots. (*b*) The Very Large Array (VLA) of the US National Radio Astronomy Observatory. This is the largest aperture synthesis radio telescope in the world.

were very faint and so high accuracy radio positions were needed to make a convincing association of the radio source with a distant galaxy. By the early 1960s, much improved positions for these extragalactic radio sources became available and many distant radio galaxies were found by searching close to the radio source positions on high quality optical photographs. In 1960, the most distant galaxy known at that time, the radio galaxy 3C 295, was discovered by Rudolf Minkowski by this technique. This procedure for finding distant radio galaxies is still one of the most important tools for cosmology – the most distant galaxies we know of, which emitted their light when the Universe was less than about one fifth of its present age, have been discovered by the radio identification technique.

In 1960, it was found that, using accurate radio positions, a few of the radio sources seemed to be associated with stars, rather than with galaxies. The optical spectra of these objects contained prominent emission lines but, unlike the spectra of stars, these lines could not be associated with lines of any of the common elements. It was not until 1962 that Maarten Schmidt of the California Institute of Technology realised that the reason the spectra were unintelligible was that the whole spectrum had been shifted to the red end of the optical spectrum. By shifting the observed spectrum back to shorter wavelengths, he was able to identify the Balmer series of hydrogen in the spectrum of the quasar 3C 273. We will discuss the full significance of the redshift of the spectrum in the next chapter – for our present purposes, we note that the shift of the spectrum to longer wavelengths is known as the *redshift* of the object and it is a measure of the velocity of recession of the object from our Galaxy. In the expanding Universe, this velocity is proportional to the distance of the object from us. Very soon after this discovery, even greater velocities of recession were measured for other star-like objects. What was remarkable about these discoveries was that these objects were as distant as the most distant galaxies known at that time. They could not, in any sense, be any normal type of star.

The remarkable object 3C 273 was the first of the *quasi-stellar objects*, or *quasars*, to be discovered (Figure 1.14). The light from the nucleus outshines the light from the galaxy by a factor of about 1000. In Figure 1.14, galaxies at the same distance as the quasar are the very faint smudges seen towards the bottom of the picture and any one of these would be totally undetectable, if it were the host galaxy for such an active galactic nucleus – the light of the galaxy would be swamped by the light of the nucleus. It can also be seen in Figure 1.14 that there is a jet pointing away from the quasar, not unlike those observed in radio sources such as Cygnus A at radio wavelengths (Figure 3.3(*a*)).

The discovery of quasars was remarkable enough, but what made matters really exciting was the fact that their optical emission was found to vary in intensity on remarkably short time-scales. In Figure 3.4, the time variations of the intensity of radiation of two active galactic nuclei are shown, in one case the optical variability of the quasar 3C 345, and in the other the variability of the Seyfert galaxy MCG-6-30-15 at X-ray wavelengths. It is apparent that, in these cases, variability can occur on time-scales as short as a day or even less.

Why are these such important observations? Suppose a source of radiation has physical size D and that it lights up very briefly. For simplicity, let us assume that the source is spherical and that D is the diameter of the sphere. What will be observed by a distant observer? Light from the front side of the source, nearest the observer, will arrive first and the light from the far side will arrive later by time t, which is just the time it takes light to cross the

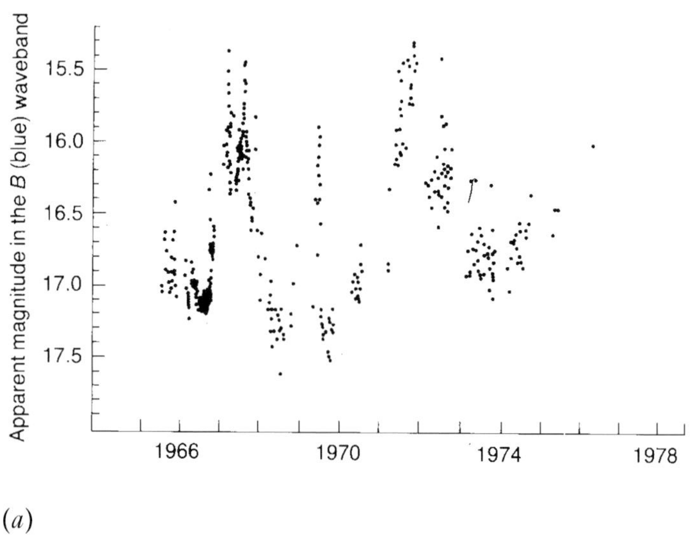

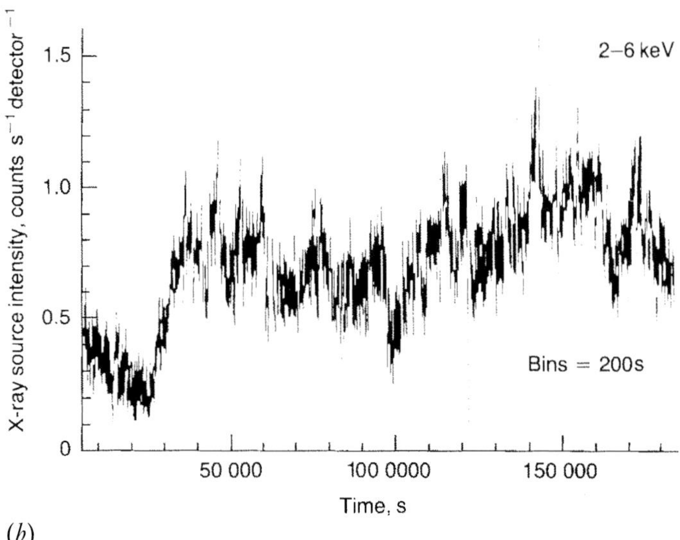

Figure 3.4. Examples of the time variability of the emission of active galactic nuclei. (*a*) The optical variability of the radio quasar 3C 345 in the blue waveband. (*b*) The X-ray variability of the Seyfert galaxy MCG-6-30-15 in the 2 to 6 keV waveband.

source, that is, $t = D/c$ where c is the speed of light. Notice that the burst of radiation can be of negligibly short duration, but the finite dimensions of the source region and the fact that the speed of light is a constant, mean that a pulse of duration roughly D/c is observed. Therefore, if a source is observed to vary significantly in time t, we know that the emission must have originated from a region of physical size *smaller* than approximately $D = ct$.

This is an example of what is known as *causality* in relativity. If we observe a source varying significantly in intensity in, say, one day, then we know that the region from which the emission originated cannot be any greater than one light day in size. We can therefore obtain good limits to the sizes of the regions from which the intense emission originates in active galactic nuclei from their time variability. These regions must be less than a light year in size, which is very tiny indeed compared with the sizes of galaxies. In fact, the enormous luminosities of quasars must originate within the very central regions of the parent galaxies – what are referred to as the galactic nuclei. The quasars must therefore contain a completely new type of energy source, in which enormous fluxes of optical, radio and X-ray emission are generated within very compact regions. We have to imagine a luminosity corresponding to about 1000 times the total luminosity of our own Galaxy coming from a region which must be smaller than the typical distance between the stars in our Galaxy and often very much smaller.

We should enter one caveat about the line of reasoning described above. These limits to the physical sizes of variable sources are good, provided the sources are not moving at very great speeds, that is, that they are not moving, or expanding, at speeds close to the speed of light. We will find that there is strong evidence that some of the most luminous sources are indeed moving at speeds close to the speed of light (Section 3.7). These considerations somewhat alter the sizes and time-scales associated with active galactic nuclei but not to such an extent that they invalidate the general line of reasoning presented above.

In many ways, the quasars were discovered too soon. Since their discovery in the early 1960s, many more examples of active galactic nuclei have been discovered and most of them are not nearly as extreme as the quasars. For example, the galaxy NGC 4151, shown in Figures

1.12 and 1.13, is a relatively nearby galaxy and contains an active nucleus with many features in common with the quasars – the nucleus is very luminous at optical and X-ray wavelengths and its intensity is variable over a period of a few days. It is probable that there are even smaller-scale versions of active nuclei, sometimes called mini-quasars, in ordinary galaxies such as our own. The present consensus is that active galactic nuclei are probably present in all galaxies. It is somewhat reassuring that the most extreme examples, the quasars, are very rare objects indeed, but they also pose the severest astrophysical problems.

When they were first discovered, the quasars posed many very difficult problems for theorists. A few astronomers found their properties so extreme that they proposed that the redshifts of the quasars might not be due to the expansion of the Universe. Within ten years of their discovery, however, it became clear how all types of active galactic nuclei could be understood in terms of energetic processes taking place in the vicinity of massive black holes. The clues to unravelling the physics of active galactic nuclei came from two quite unexpected discoveries – the radio pulsars and the binary X-ray sources.

3.3 The discovery of pulsars

The 1960s and early 1970s were amazing years for astronomy, astrophysics and cosmology. One discovery succeeded another with bewildering rapidity. For astrophysics as a whole, no discoveries were more important than those of pulsars in 1967 and binary X-ray sources in 1971. It might seem perverse to introduce these stellar objects into the middle of a discussion of the physics of quasars but they provide crucial clues about the processes of energy generation in compact objects. These ideas can be applied with some degree of success to the physics of the most extreme active galactic nuclei.

The radio pulsars came as more or less a complete surprise when they were discovered by Antony Hewish and Jocelyn Bell(-Burnell) in 1967. The story had begun several years earlier when Hewish had pioneered the study of the 'scintillation' of radio sources at low radio frequencies. Just as the stars are observed to twinkle, so the observed intensities of radio sources flicker or 'scintillate' when observed at low radio frequencies. By 1964, Hewish had established that the fluctuating intensities of compact radio sources at low radio frequencies are due to the scattering of the radio waves by irregularities in the diffuse gas flowing out from the Sun, known as the *solar wind*. He and his colleagues established that this phenomenon, known as interplanetary scintillation, could be used to study both the properties of the solar wind and the structures of the radio sources. The strongest scintillating radio sources were the most compact sources and these were very frequently quasars.

In 1964, Hewish was awarded a grant to design and build a large low-frequency array of dipole antennae with which to carry out detailed studies of all aspects of interplanetary scintillation. To obtain high sensitivity at the low frequency of 81.5 MHz (3.7 m wavelength), the array had to be very large, 1.8 hectares (4.5 acres) in area (Figure 3.5(*c*)). This was the key technical development, since the scintillations of the sources had to be recorded on a timescale of one-tenth of a second. Jocelyn Bell joined the programme in 1965 as Hewish's research student. The first sky surveys were carried out in July 1967 and Bell soon discovered a strange source which seemed to consist entirely of scintillating radio signals (Figure 3.5(*a*)). The source was not always present and its nature remained a mystery. In November 1967, the source was observed with a recorder with a shorter time-constant and it was found that the

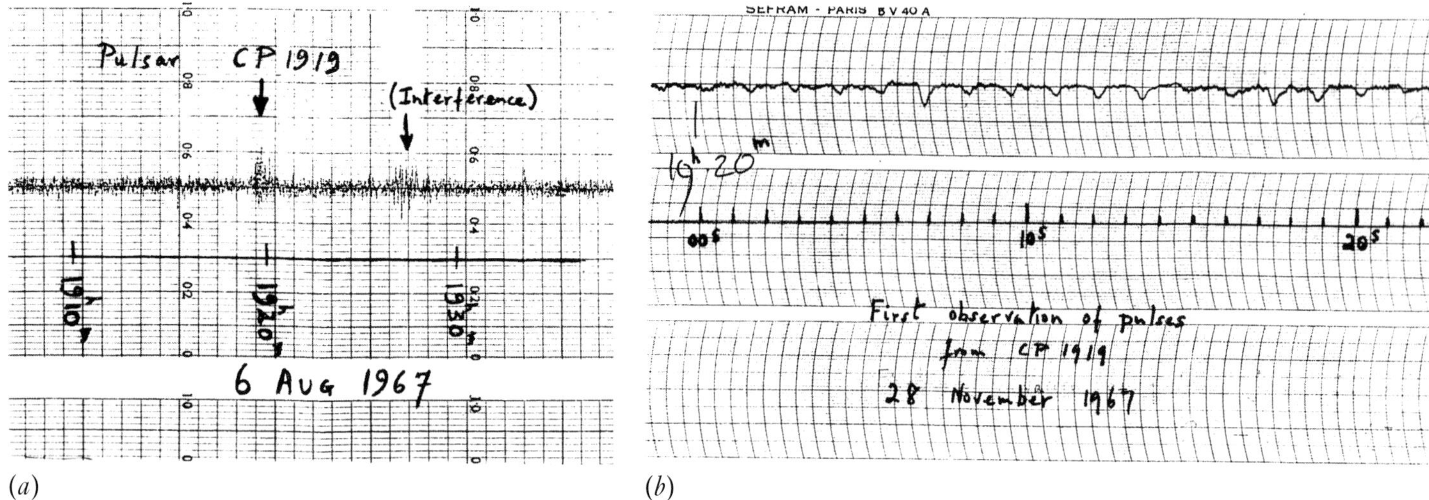

(*a*) (*b*)

(*c*)

Figure 3.5. The discovery records of the first pulsar, PSR 1919 + 21. (*a*) The first records of the strange scintillating source, labelled CP 1919. Note the subtle differences between the signal from the source and the neighbouring signal due to terrestrial interference. (*b*) The signals from PSR 1919+21 observed with a recorder with a shorter time-constant than the discovery record, showing that the signal consisted entirely of regularly spaced pulses with period 1.33 seconds. (*c*) The Cambridge 4-acre array built by A. Hewish and his colleagues, with which the radio pulsars were discovered. The photograph shows Jocelyn Bell (-Burnell) in front of a small section of the phased array.

source consisted entirely of a series of regular pulses, the interval between them being about 1.33 seconds (Figure 3.5(*b*)). Over the next few months, three further examples of these pulsating sources were discovered, one of them with a pulse period of only a quarter of a second. It was not long before the name *pulsating radio source* was abbreviated to *pulsar*. In the early days, the four sources were called LGM1, LGM2, LGM3 and LGM4, LGM standing for 'Little Green Man', because the discovery of what seemed to be some sort of celestial Morse code was totally unexpected and it was just conceivable that the pulses could have been messages from extraterrestrials. I remember admiring the Christmas card which Jocelyn Bell sent to Antony Hewish at Christmas 1967 – it consisted entirely of a series of pulses, which, when decoded, contained a Christmas greeting.

It was soon established that the only types of star which could produce such regular pulses with periods of less than a second were the *neutron stars*. Figure 3.6 is a sketch of a rotating, magnetised neutron star, indicating how it can produce pulsed radio emission. Neutron stars are very compact stars indeed and, in the case of pulsars, they possess very strong magnetic fields. It is assumed that the magnetised neutron star emits a beam of radio emission along the poles of the dipole magnetic field. To create a pulsar, the axis of the magnetic dipole has to be misaligned with respect to the rotation axis of the neutron star. Thus, the poles of the magnet are swept around at the rotation period of the neutron star. This means that the neutron star must rotate at a frequency of about once or more per second, a quite extraordinary rotation rate for an object which is as massive as the Sun. This speed of rotation can only conceivably occur in extremely compact stars such as neutron stars, because larger stars would be ripped apart by centrifugal forces.

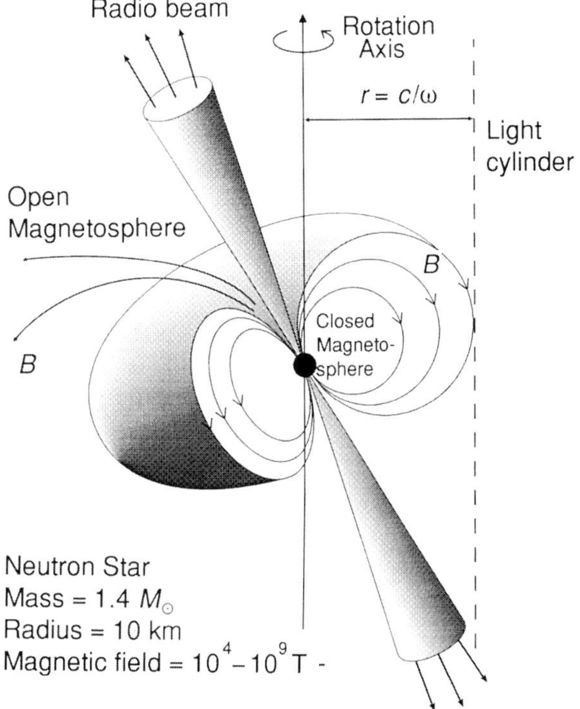

Figure 3.6. A schematic model of a pulsar as a rotating, magnetised neutron star, in which the magnetic and rotation axes are misaligned. The radio emission is beamed along the magnetic poles of the neutron star and consequently, at a distance, a bright pulse of radio emission is observed once per rotation of the neutron star, if the beam intersects the observer's line of sight. Typical properties of neutron stars are shown on the diagram. The magnetic field is measured in tesla (T); for reference, the Earth's magnetic field is about 0.00005 T.

What are neutron stars? We need to take up the story of the death of stars where we left it in the last chapter. Once a star has burned up all its available nuclear fuel, it no longer has a source of energy to provide the thermal pressure needed to resist the pull of gravity – consequently, the core of the star collapses. The only source of pressure left is the quantum mechanical pressure due to the fact that particles such as protons, neutrons and electrons are not permitted to occupy precisely the same quantum mechanical state. This is a purely quantum mechanical phenomenon and has no counterpart in classical physics. It is this same rule which prevents electrons occupying exactly the same orbits within atoms and which leads to the enormous diversity of phenomena found in chemistry. This law of quantum mechanics prevents particles coming too close together and the pressure associated with this repulsive force is known as *degeneracy pressure*.

White dwarf stars can be convincingly associated with the end point of evolution of stars like the Sun and, in these, the electrons are responsible for providing the degeneracy pressure which enables them to resist gravitational collapse. The theory of these stars was worked out as soon as Pauli's exclusion principle and the concept of degeneracy pressure were discovered in the 1920s. White dwarfs are compact stars, having radii only about one-hundredth the radius of the Sun and they have no internal energy sources. They are therefore very faint stars and three of them can be seen lying to the bottom left of the main sequence in the Hertzsprung–Russell diagram shown in Figure 2.9(*a*). There must be large numbers of them in our own Galaxy and they form naturally as the end points of the evolution of stars with mass roughly the mass of the Sun, after they have evolved to the tip of the giant branch. At that point, the stars become unstable and eject their envelopes non-catastrophically to form what are known as *planetary nebulae* (Figure 2.12). At the same time, their cores contract until electron degeneracy pressure prevents them collapsing any further. These compact remnants cool to become white dwarfs, as indicated by the evolutionary tracks shown on Figure 2.13(*b*).

There is, however, a yet more compact final state for stars, in which the degeneracy pressure of neutrons prevents the star collapsing under gravity. It may seen remarkable that neutrons can provide pressure support for a star, since it is well known that free neutrons decay into protons, electrons and neutrinos with a half-life of only about 11 minutes. What happens is that, if the gas becomes very dense, the electron degeneracy pressure becomes so great that the electrons attain very high energies, in fact, relativistic energies. When the energy of the electrons becomes greater than 1.29 MeV, they can interact with protons to form neutrons by the reaction

$$p + e^- \rightarrow n + \nu_e.$$

Normally, the neutrons would decay but the gas is highly degenerate and so there are no states available for the electrons, if they were to be emitted. Thus, the degeneracy of the dense gas stabilises the neutrons against decay. This process only takes place at densities greater than about 10^{10} kg m^{-3}, that is, about 10 000 000 times the density of water.

As a result, neutron stars are very compact objects indeed. The radius of a typical neutron star is about 10 to 15 km and its mass is roughly the mass of the Sun. The density of the material inside the star is about 10^{17} to 10^{18} kg m^{-3}. Similar densities are only found inside the nuclei of atoms. Indeed, neutron stars can be thought of as being giant nuclei, each consisting of about 10^{60} nucleons – all 10^{60} nucleons are packed tightly up against one another and are held there by the force of gravity. The physical conditions inside neutron stars are thus extreme and unlike anything encountered in laboratory physics. Figure 3.7 provides some impression of the remarkable physical processes which take place in their interiors.

Remarkably, the existence of neutron stars had first been proposed by Walter Baade and Fritz Zwicky in 1934, within a couple of years of the discovery of the neutron by James Chadwick. They proposed that neutron stars would be formed in the explosions of stars at the ends of their lifetimes and that they might be the source of the high energy particles

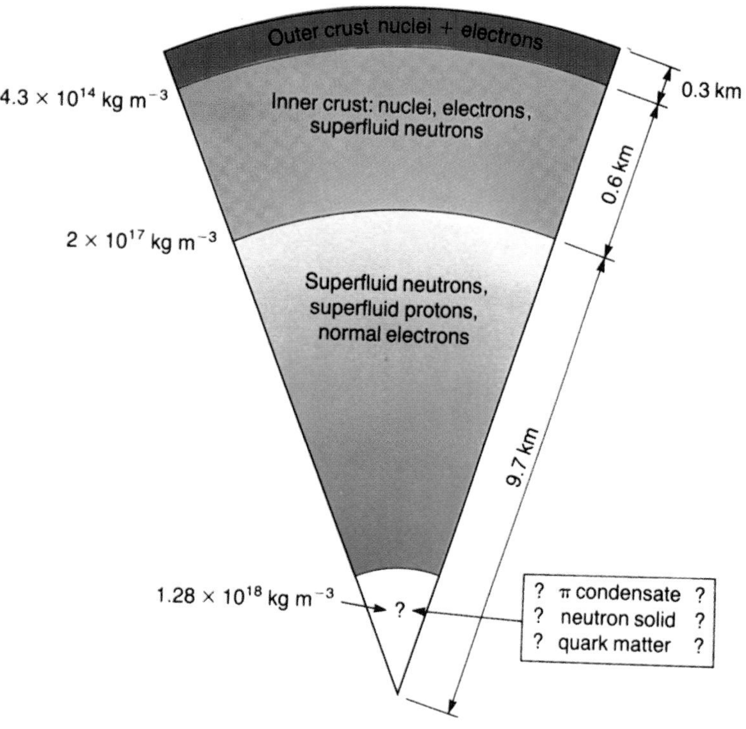

Figure 3.7. A representative model showing the internal structure of a neutron star of mass 1.4 times the mass of the Sun.

known as *cosmic rays*. It is worth allowing Zwicky to describe this prescient suggestion in his own words:

> In the *Los Angeles Times* of January 19, 1934, there appeared an insert in one of the comic strips, entitled 'Be Scientific with Ol'Doc Dabble' quoting me as having stated:
>
> > 'Cosmic rays are caused by exploding stars which burn with a fire equal to 100 million suns and then shrivel from $\frac{1}{2}$ million miles diameter to little spheres 14 miles thick', says Prof. Fritz Zwicky, Swiss Physicist.
>
> This, in all modesty, I claim to be one of the most concise triple predictions ever made in science. More than thirty years were to pass before the statement was proved to be true in every respect.

Zwicky was an extraordinary character and he must be given due credit for his remarkable insights. The formation of neutron stars in supernova explosions was fully confirmed by the discovery of young radio pulsars in the Crab and Vela supernova remnants. It is generally believed that neutron stars are formed in the explosions of massive stars and the statistics of pulsars are roughly consistent with this picture. To complete Zwicky's list of successes, it has been found that supernova remnants are among the most intense Galactic radio sources. For example, the radio image of the supernova remnant Cassiopeia A (Figure 3.8) is more or less identical to the map of its X-ray emission (Figure 1.25), but the radio emission is due to the synchrotron radiation of very high energy electrons. Therefore, particles of cosmic ray energies can certainly be accelerated in supernova remnants. The most convincing theory for the origin of the flux of cosmic ray particles observed in the interstellar medium by both their radio and γ-ray emission is that they were accelerated in supernova explosions, such as Cassiopeia A.

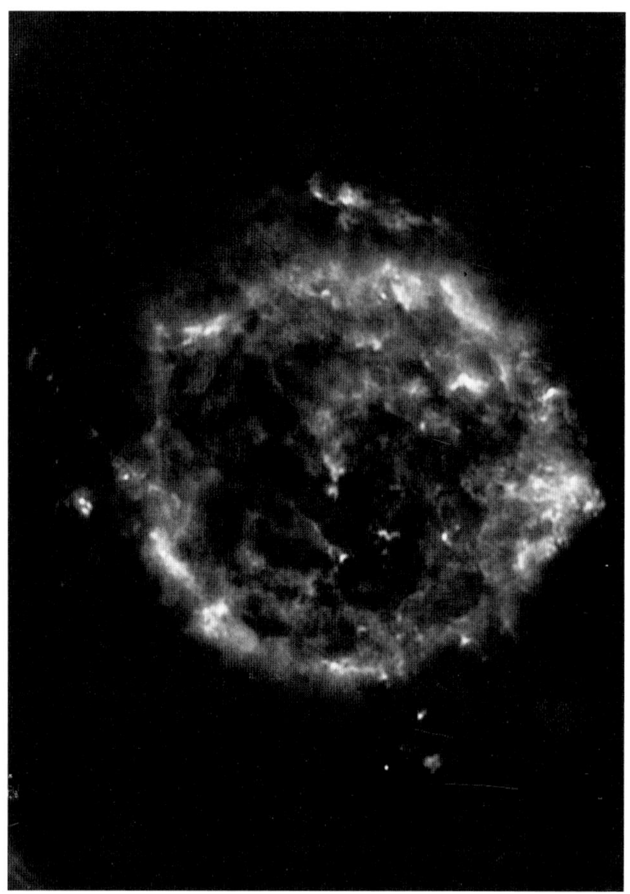

Figure 3.8. A radio image of the supernova remnant Cassiopeia A observed by the Ryle Telescope at Cambridge. The radio emission is the synchrotron radiation of ultrarelativistic electrons and these have been accelerated in the supernova remnant. This image should be compared with the optical and X-ray images of the same source (Figures 1.24 and 1.25 respectively).

3.4 The supernova SN1987A

Direct evidence for the processes which take place when massive stars die has been provided by the supernova which exploded in the Large Magellanic Cloud in February 1987, the supernova known as SN1987A (Figure 3.9). This has been one of the most exciting and important events in astronomy this century. It was the brightest supernova since Kepler's supernova of 1604 and was the first bright supernova to be observed in detail with the full power of modern instrumentation. The supernova coincided precisely with the position of a bright blue supergiant star known as Sanduleak –69 202, which disappeared following the supernova explosion. Models of the supernova explosion have indicated that the star must have had mass about 20 times the mass of the Sun.

One of the pieces of great good fortune was that, at the time of the explosion, neutrino detectors were in operation at the Kamiokande experiment in Japan and at the Irvine–Michigan–Brookhaven (IMB) experiment located in an Ohio saltmine in the USA. Both experiments were designed for an entirely different purpose, which was to search for evidence of proton decay. Remarkably, a brief burst of neutrinos was detected by both experiments just before the supernova was observed on 24 February 1987. Twelve neutrinos were detected by the Kamiokande experiment and 8 by the IMB experiment and the burst of neutrinos was detected simultaneously at the two detectors. All 20 neutrinos arrived within an interval of 12 seconds. The supernova was only observed optically several hours after the

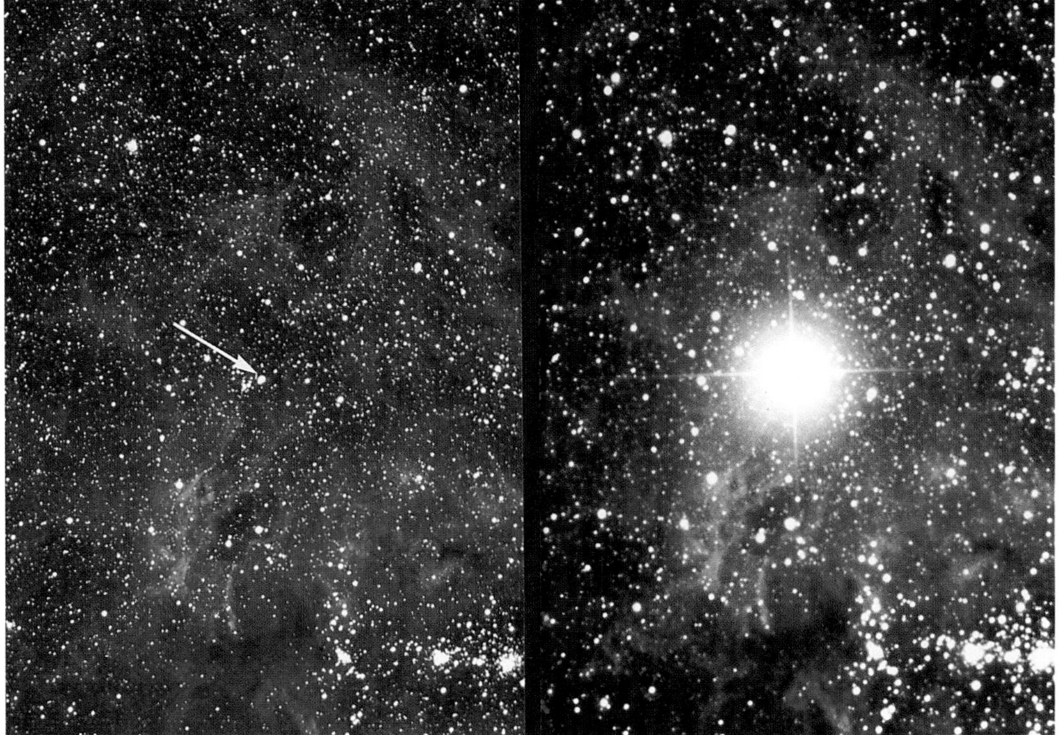

Figure 3.9. The field of the supernova SN1987A in the Large Magellanic Cloud, before and after the explosion which was first observed on 24 February 1987.

pulse of neutrinos had arrived and this is entirely consistent with a picture in which the neutrinos escaped more or less directly from their point of origin in the collapsed core of the presupernova star whereas the optical light had to diffuse out through the supernova envelope.

The observation of the neutrino flux from the supernova is uniquely important for the theory of stellar evolution. The supernova was ideally placed from the astronomical point of view because the distance of the Large Magellanic Cloud is accurately known and so the neutrino luminosity of the supernova can be found. It turns out that the neutrino luminosity and the energies of the neutrinos from SN1987A are precisely what would be expected when a neutron star forms. Although no pulsar has yet been detected in the remnant of that supernova, the flux of neutrinos emitted when the progenitor star collapsed was exactly what would have been expected if a neutron star had formed. These observations provide strong support for the essential correctness of our understanding of the late stages of stellar evolution.

Another beautiful set of observations of the supernova concerns the rate at which the light from the supernova declined and the formation of the heavy elements. One of the remarkable features of this supernova, which is common to most supernovae, is that, after the initial outburst, the light from the supernova declined exponentially with a half-life of about 77 days (Figure 3.10). The most promising theory to account for this decay is that, in the collapse of the core of the progenitor of the supernova, the synthesis of the heavy elements proceeded all the way through to iron and that, in addition to the stable forms of iron and similar iron-group metals, unstable nuclei were created as well. These were ejected along with the other synthesised elements in the supernova explosion. The most important radioactive chain involves the decay of the radioactive isotope of nickel ^{56}Ni to radioactive cobalt ^{56}Co, which then decays to

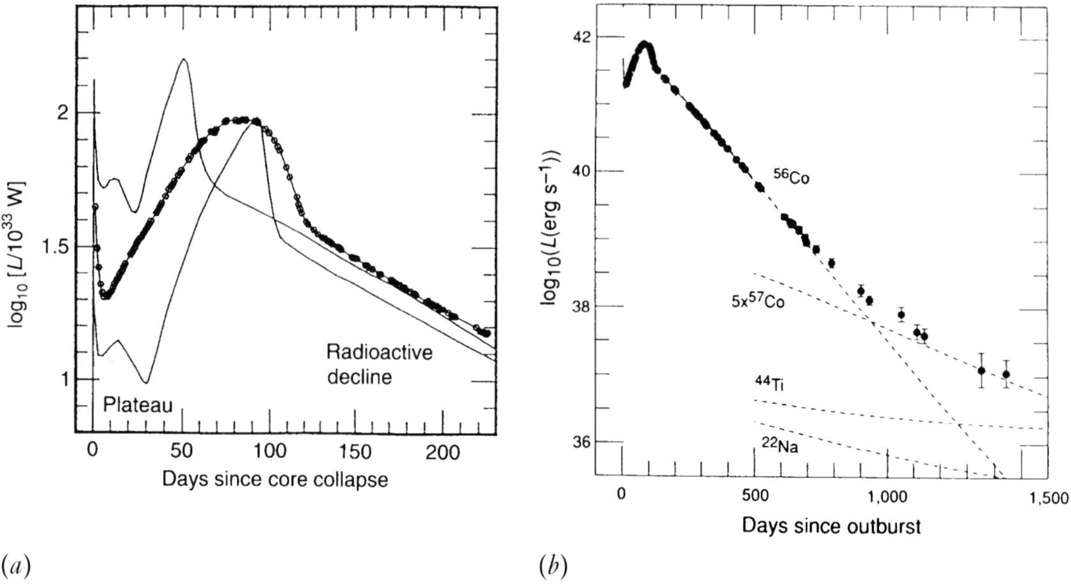

Figure 3.10. The light curve of the supernova SN1987A during its first five years. This light curve shows how the total luminosity of the supernova at optical, ultraviolet and infrared wavelengths declined during the first five years since the explosion in February 1987. In (*a*), the thin lines represent some theoretical models of the early light curve of the supernova. In (*b*), the dashed lines show how the luminosity is expected to decline as the different radioactive species come into play. The energies deposited by the radioactive nuclides are based upon the following initial masses: 0.075 $M_\odot$ of ^{56}Ni, 10^{-4} $M_\odot$ of ^{44}Ti, 2×10^{-6} $M_\odot$ of ^{22}Na and 0.009 $M_\odot$ of ^{57}Co, the last corresponding to five times the solar abundances of ^{57}Fe/^{56}Fe.

the stable isotope of iron ^{56}Fe. The first decay has a half-life of only 6.1 days, while the second decay, involving the decay of ^{56}Co, has exactly the required half-life of 77.1 days to account for the decline of the optical luminosity of the supernova. If this is the correct explanation of the light curve of the supernova, about 0.07 $M_\odot$ of ^{56}Ni must have been created in the supernova, in good agreement with the expectations of theoretical models of the process of nucleosynthesis.

A consequence of this model is that, as the supernova declines, the optical spectrum of the supernova should change. Absorption lines associated with the radioactive species ^{56}Ni and ^{56}Co should decrease in intensity relative to the lines of ^{56}Fe as the radioactive isotopes decay. These changes in the relative strengths of the lines of nickel and cobalt have now been observed. The totality of these observations provides direct confirmation for the radioactive theory of the origin of the supernova light curve, for the formation of the iron-group elements in supernova explosions and for the ejection of the processed material into the interstellar medium.

3.5 The discovery of X-ray binary sources

The discovery of neutron stars as the parent bodies of pulsars was a piece of great good fortune and resulted from the exploration of an area of parameter space, which had not previously been accessible to astronomers, namely short-time-scale astronomy at long radio wavelengths. The discovery of pulsars opened up totally new perspectives for astronomy. The neutron stars had been largely theoretical constructs up till 1968 and there had been little prospect of detecting them observationally because they are such tiny objects. One possibility had been that, if the surfaces of the neutron stars were very hot, they might be detectable as X-ray sources, but X-ray astronomy was still in its infancy.

The next key discovery was provided by the emerging discipline of X-ray astronomy in 1971, but not quite in the manner which might have been expected. Up till 1970, all X-ray astronomy was carried out from rockets, which allowed the X-ray telescopes to view the sky for about five minutes while the rocket was above the Earth's atmosphere, before the rocket and its payload crashed back to Earth. These rocket experiments showed that there exist

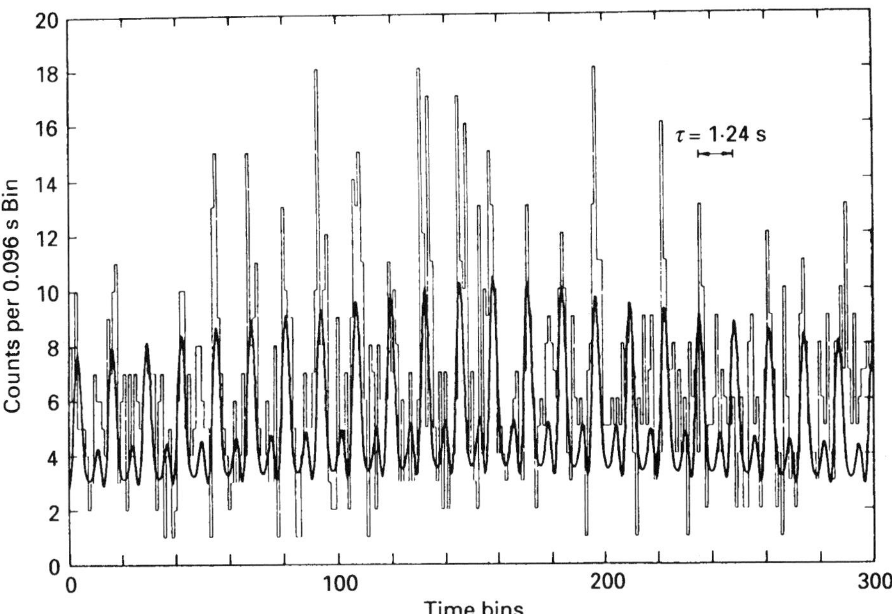

Figure 3.11. The discovery record of the pulsating X-ray source Hercules X-1 (Her X-1) observed by the Uhuru X-ray satellite in 1971. The X-rays are emitted in regular pulses with period about one second, similar to the periods found among the radio pulsars.

highly variable X-ray sources. During the 1960s, X-ray astronomers were bemused by the fact that the X-ray sources would come and go, they would appear and disappear and it was difficult to establish whether they were real or not. The Uhuru X-ray satellite, launched in 1970, was the first satellite dedicated to X-ray astronomy and it clarified many of the problems of the 1960s. One of the most important discoveries of the mission was that, among the Galactic X-ray sources, there are sources which pulsate very rapidly at X-ray wavelengths. Figure 3.11 shows the discovery record of the pulsations in the source Hercules X-1. The period of the X-ray pulsations is very short and similar to those of the radio pulsars. This strongly suggested the presence of a neutron star in the X-ray source. The process by which the X-rays are emitted is, however, completely different.

The sources would not be X-ray emitters unless there is gas present at a temperature of about 10 000 000 K or more, as can be appreciated from Figure 1.20. An important clue came from the fact that many of the pulsating X-ray sources were found to be members of binary star systems. In these, one of the stars, the primary star, is a normal star but the secondary companion star is completely invisible optically. In some cases, the X-ray source disappeared regularly for a day or so and then reappeared, with the same period as the binary orbit. This provided direct evidence for the eclipse of the X-ray source by the primary star, which occurs when the plane of the binary orbit is roughly perpendicular to the plane of the sky.

As soon as the binary nature of these X-ray sources was established, it did not take the theorists long to work out that there is a simple means of creating hot gas in these systems. If the binary star system consists of a normal star and a neutron star and matter is transferred from the normal star onto the compact star, the matter is heated to a very high temperature when it falls onto the surface of the compact star. This process is known as *accretion* and, in one way or another, is responsible for the intense X-ray emission from binary X-ray sources. The matter dragged off the companion star onto the compact star is channelled by the magnetic field onto the magnetic poles of the compact star and so there must be very hot regions

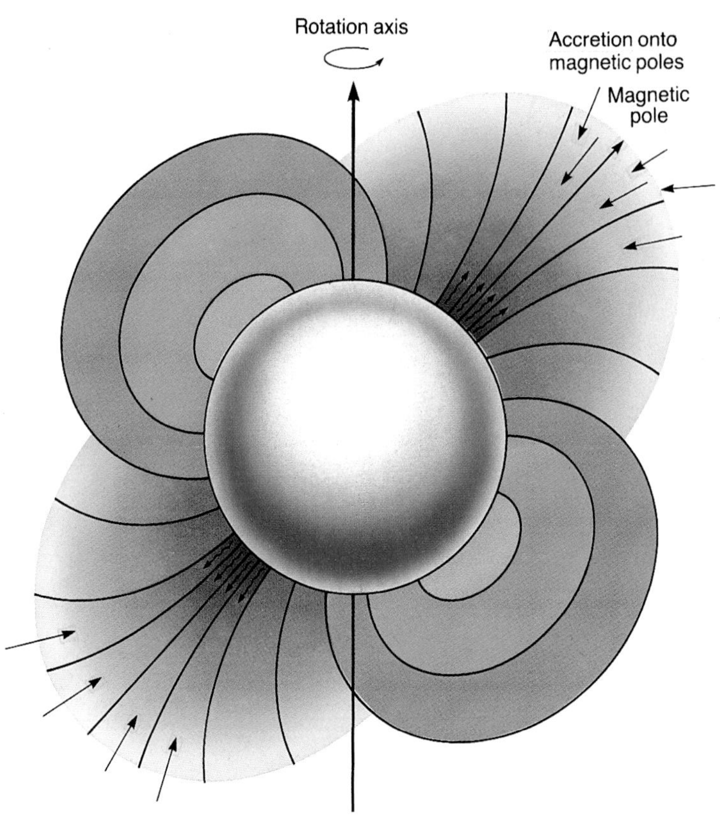

Figure 3.12. A schematic model illustrating how accretion onto a neutron star with a magnetic field non-aligned with the rotation axis can give rise to an accretion column. The rotation of the neutron star gives rise to pulsed X-ray emission when the source is observed at a distance. The accreting matter is channeled down the magnetic field lines onto the magnetic poles of the neutron star.

close to the magnetic poles of the rotating neutron star. As these hot spots are swept round by the rotation of the neutron star, pulses of X-ray emission are observed at the rotation frequency of the neutron star (Figure 3.12). As we will show, these studies of pulsars and X-ray binaries provide a number of important clues for understanding how energy can be generated at the enormous rates needed to power the most active galactic nuclei.

There is, however, a major problem – neutron stars are only stable if their masses are less than about twice the mass of the Sun. The same is true for white dwarfs. Good mass estimates have now been made for a number of neutron stars, which are members of binary systems, and they all turn out to have masses about 1.4 times the mass of the Sun. This is very good news and in good agreement with theory. We still have to address the question, however, of what happens if the dead remnants have mass greater than twice the mass of the Sun. There turns out to be no physical force which can prevent the gravitational collapse of such objects to black holes. Our next task is to study black holes and, to do this, we need to discuss the general theory of relativity, the relativistic theory of gravity.

3.6 General relativity and black holes

The first discussion of what we would now call black holes was presented by the Reverend John Michell in a paper read before the Royal Society of London in 1783, more than 130 years before Einstein's general theory of relativity. In modern parlance, Michell asked the question, 'Suppose a star has a mass M. What would the radius of that star be if the escape velocity from its surface were equal to the speed of light?' The *escape velocity* is the speed a rocket has to have at the surface of the Earth if it is to escape from the pull of the Earth's gravity. In other words, it is the velocity which enables the rocket to travel an infinite distance from the Earth and to have zero velocity when it gets there. If the rocket had initial velocity less than the escape velocity, it would fall back to Earth. Michell asked exactly the same question, but now the escape velocity from the star is to equal the speed of light. In his own words,

> ... all light emitted from such a body would be made to return towards it, by its own proper gravity.

Remarkably, he found precisely the same answer as is derived in the relativistic version of the same problem – no light can escape from within a certain radius.

Michell's studies were made long before the discovery of the special and general theories of relativity, which are the modern tools used to tackle problems in which the speeds of objects approach the speed of light. Einstein discovered not one but two theories of relativity. The *special theory of relativity* was discovered in 1905 and describes how Newton's laws of motion are modified when bodies move at high speeds. Among the implications of the theory is the equivalence of mass and energy, $E = mc^2$, which we discussed in Section 2.1. Even more fundamental is the idea that space and time are no longer the independent quantities they appear to be in everyday life. We can no longer think of time and three-dimensional space as separate entities but rather, we should think in terms of four-dimensional *space–time*. The special theory of relativity has been subjected to the most rigorous experimental tests and it has passed all of them with flying colours.

The *general theory of relativity* is the relativistic theory of gravity. Newton's law of gravity is very accurate indeed on the scale of bodies within the Solar System but it has to be modified to take account of the fact that all our theories should be consistent with the special theory of relativity. The discovery of the general theory of relativity proved to be a much more difficult task than that of the special theory and it took many years of very hard work before Einstein published the completed theory in 1915. The theory has to take account of the facts that we live in space–time, and not in space and time, and also that gravitational fields influence the paths of light rays, as was appreciated by John Michell. The essence of the theory can be summed up in the two statements:

- Matter bends space–time.
- Matter moves along paths in bent space–time.

Einstein's great achievement was to develop a mathematical theory which incorporated both features in a self-consistent manner. This is precisely the type of theory needed to understand the behaviour of matter in the presence of the very strong gravitational fields of neutron stars and black holes.

Before applying the theory to this problem, we should ask how far we can trust the theory. Most of the tests of general relativity involve the observation of astronomical objects and there has been excellent progress in testing the predictions of the theory, for example, by measuring the deflection of electromagnetic signals from distant astronomical objects as they are occulted by the Sun, and the 'fourth' test of general relativity, discovered by Irwin Shapiro,

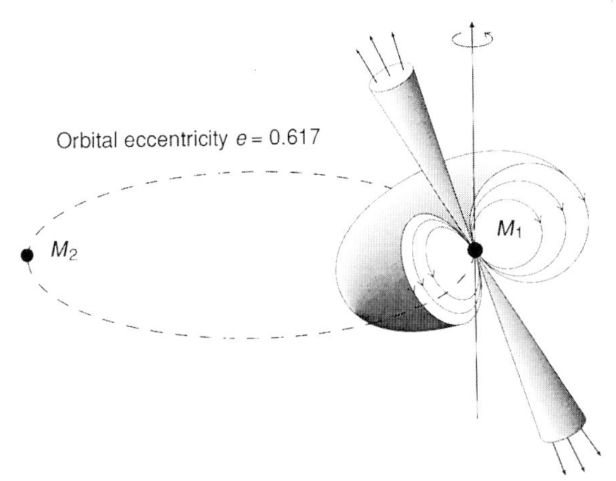

Binary period = 7.751939337 hours
Pulsar period = 59 milliseconds
Neutron star mass M_1 = 1.4411(7) $M_\odot$
Neutron star mass M_2 = 1.3874(7) $M_\odot$

Figure 3.13. A schematic representation of the binary pulsar PSR 1913+16. As a result of the ability to measure precisely many parameters of the binary orbit by ultra-precise pulsar timing, the masses of the two neutron stars have been measured with very high precision.

of the time delay as the electromagnetic radiation from a distant object passes through the gravitational field of a massive body such as the Sun.

The most spectacular results have come, however, from radio observations of radio pulsars. Joseph Taylor and his colleagues have used the Arecibo radio telescope to demonstrate that pulsars are the most stable clocks we have available in the Universe. In their observations of some of the most stable pulsars, they have found variations in the time-keeping of the most accurate laboratory clocks, by comparing the times measured by two pulsars with that of a standard clock.

Among the most intriguing systems for testing the general theory of relativity are those pulsars which are members of binary systems. More than 20 of these are now known, the most important being those in which the other member of the binary system is also a neutron star and in which the neutron star pair forms a close binary system. The first of these was discovered by Russell Hulse and Joseph Taylor in 1974 and is known as the pulsar PSR 1913+16 – it is illustrated schematically in Figure 3.13. The system has a binary period of only 7.75 hours and the orbital eccentricity is large, $e = 0.617$. This system is a pure gift for the relativist. To test general relativity, we require a perfect clock in a rotating frame of reference and systems such as PSR 1913+16 are ideal for this purpose. The neutron stars are so inert and compact that the binary system is very 'clean' and so can be used for some of the most sensitive tests of general relativity yet devised.

In Figure 3.14, the determination of the masses of the two neutron stars in the binary system PSR 1913+16 is shown, assuming that general relativity is the correct theory of gravity. Various parameters of the binary orbit, which depend upon M_1 and M_2, can be measured very precisely and these provide different estimates of functions involving the masses of the two neutron stars. In Figure 3.14, various parameters of the binary orbit are shown, those which have been measured with very good accuracy being indicated by an asterisk. It can be seen that the different loci intersect precisely at a single point on the $M_1 - M_2$ diagram. Some

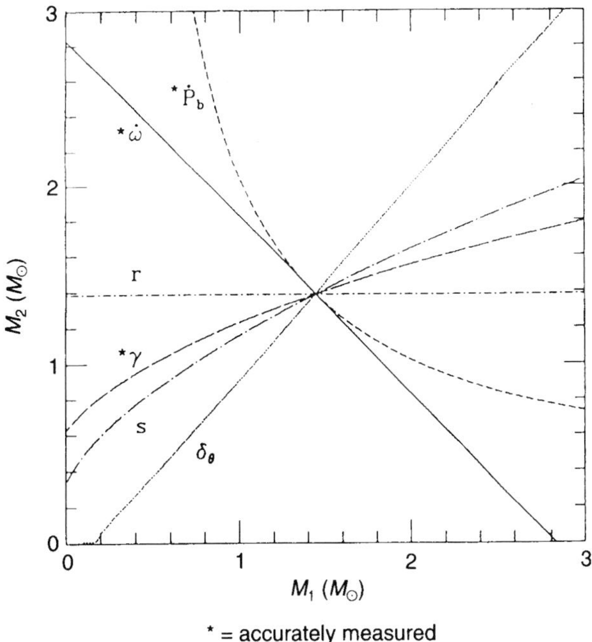

Figure 3.14. The measurement of the masses of the neutron stars in the binary system PSR 1913+16 from very precise timing of the arrival times of the pulses at the Earth. Different parameters of the neutron star's orbit depend upon different combinations of the masses M_1 and M_2 of the neutron stars. It can be seen that the lines intersect very precisely at a single point in the M_1 M_2 diagram.

measure of the precision with which the theory is known to be correct can be obtained from the accuracy with which the masses of the neutron stars are known, as indicated in Figure 3.13.

A second remarkable observation has been the measurement of the rate of loss of orbital rotational energy of the neutron star pair by the emission of gravitational waves. These waves are the gravitational equivalent of electromagnetic waves – when electric charges are accelerated, they radiate electromagnetic radiation and, in a similar way, when masses are accelerated relative to one another under the influence of gravity in an appropriate way, they emit gravitational waves. To date, gravitational waves have not been detected directly from any astronomical object, because of the lack of sensitivity of the present generation of gravitational wave detectors. A close binary system loses energy by the emission of gravitational radiation and the rate at which energy is lost can be predicted precisely once the masses of the neutron stars and the parameters of the binary orbit are known. The changes in orbital phase of the binary orbit have been observed over a period of 17 years and they agree precisely with the predictions of general relativity (Figure 3.15). Thus, although the gravitational waves themselves have not been detected, exactly the correct energy loss rate from the system has been measured – this is convincing evidence for the existence of gravitational waves and this observation acts as a spur to their direct detection by future generations of gravitational wave detectors. This is a very important result, since it enables a wide range of alternative theories of gravity to be eliminated. General relativity has therefore passed every test which has been made of the theory and we can have confidence that it is an excellent description of the relativistic theory of gravity.

One final point about these tests is important for the considerations of Chapters 4 and 5. The same techniques of accurate pulsar timing can be used to determine whether or not the gravitational constant G has changed with time. These tests are slightly dependent upon the equation of state used to describe the interior of the neutron stars, but, for the complete

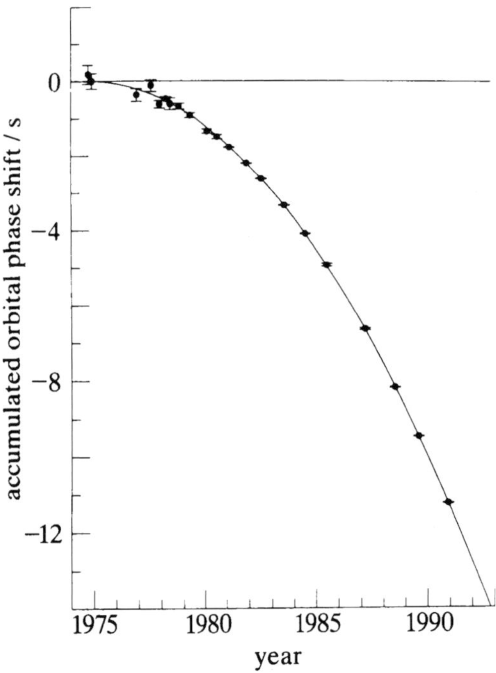

Figure 3.15. The observed change in orbital phase as a function of time for the binary neutron star system PSR 1913+16 (points with error bars) compared with the expected changes due to gravitational radiation energy loss by the binary system (solid line).

range of possible equations of state, the limits to the time variation of the gravitational constant correspond to less than about one part in 10^{11} per year. Thus, there can have been little change in the value of the gravitational constant over typical cosmological timescales, which are about 10 to 20 billion years. This result is important for the construction of cosmological models.

The properties of black holes are derived from studies of collapsed masses according to the general theory of relativity. The simplest are the *spherically symmetric* or *Schwarzschild black holes*, which are characterised solely by their masses, M. These objects are similar to the stars considered by John Michell. The radius at which the escape velocity is equal to the speed of light is known as the *Schwarzschild radius*, R_s, and is given by the simple formula

$$R_s = \frac{2GM}{c^2} = 3\left(\frac{M}{M_\odot}\right) \text{ km}$$

where G is the gravitational constant, c is the speed of light, M is the mass of the black hole and $M_\odot$ is the mass of the Sun. Thus, the Schwarzschild radius of a solar mass black hole, $M = M_\odot$, is only 3 km. The Schwarzschild radius plays a similar role to the critical radius described by John Michell – radiation can pass inwards through the Schwarzschild radius but it cannot pass from inside this radius to the outside. This is why the object is called a *black* hole – it emits no radiation, according general relativity.

The reason the object is called a *hole* is that, if matter or radiation get too close to it, they inevitably spiral into the hole, despite the fact that they may have some angular momentum. According to general relativity, when matter approaches too close to a black hole, rotation is incapable of preventing the matter from collapsing to the centre and, in fact, now helps, rather than hinders, the matter to fall in. We can think of this as being due to the fact that, according to the special theory of relativity, the rotational kinetic energy of the matter contributes to its mass and so the force of gravity acting on it is greater than in the Newtonian case. From

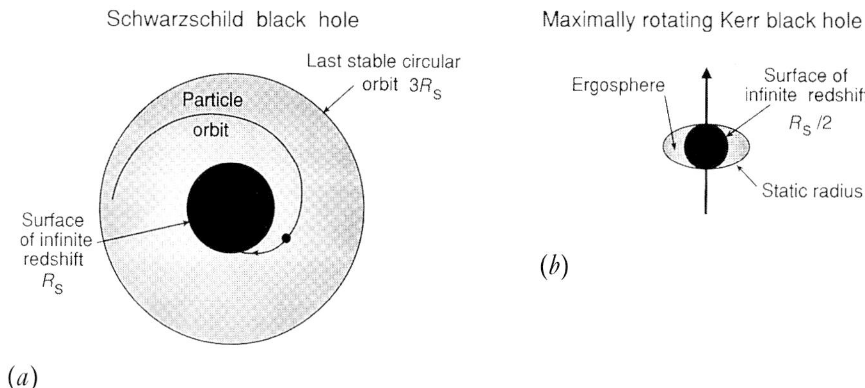

Figure 3.16. Illustrating the structures of (*a*) spherically symmetric (Schwarzschild) black holes and (*b*) rotating (Kerr) black holes. In the latter case, the black hole is rotating as fast as possible.

the astrophysical point of view, the important point is that there is a *last stable circular orbit* about a Schwarzschild black hole and it lies at a radius which is only three times the Schwarzschild radius, $r = 3R_S$. Matter can take up stable circular orbits with any radius greater than this value but, as soon as its orbit has radius $3R_s$ or less, matter is dragged inexorably into the black hole (Figure 3.16(*a*)).

I always find it remarkable that the radii of solar mass neutron stars are only about 10 km. The Schwarzschild radius of a black hole of the same mass is only 3 km and the radius of the last stable circular orbit is therefore about 9 km – in other words, the neutron stars are already almost as dense as they could possibly be without collapsing to form a black hole. Thus, in the course of stellar evolution, objects are produced which come very close indeed to becoming black holes, but are prevented from doing so by the quantum mechanical pressure of neutrons. I have always considered that this is a highly suggestive argument in favour of the existence of black holes – not all collapsing stars can have been so accurately tuned to make neutron stars rather than black holes. Furthermore, if mass is piled onto the surface of a neutron star, as occurs in X-ray binaries containing neutron stars, the mass of the neutron star will eventually exceed the limit for stability as a neutron star and so collapse to a black hole.

The other forms of black hole are the *rotating* or *Kerr black holes*. In general, it is expected that black holes should have some rotation (or, more precisely, angular momentum) because the matter out of which they formed is likely to have some net rotation, for exactly the same reasons that protostars are likely to acquire some rotation (see Section 2.5). In addition, they acquire angular momentum because the infalling matter may well have angular momentum and this speeds up the rotation of the black hole. There is, however, a maximum amount of angular momentum which a rotating black hole can possess. If it had any more, it would spin so rapidly that it would never have formed in the first place – it would have been torn apart by centrifugal forces. The upper limit to the angular momentum, $\mathcal{J}$, of a rotating black hole is $\mathcal{J} = GM^2/c$.

The rotating black holes are even more exotic than the spherically symmetric variety. There is again a 'surface of infinite redshift', from inside which radiation and matter cannot escape. In the case of maximally rotating black holes, this radius is smaller than in the non-rotating case and is equal to half of the Schwarzschild radius. Outside this surface, there is a volume in the form of an ellipsoidal shell known as the *ergosphere* of the black hole, in which particles must corotate with the black hole (Figure 3.16(*b*)). From the point of view of energy generation, the important point is that the radius of the last stable orbit about a maximally rotating black hole is equal to the radius of the surface of infinite redshift, which is only $0.5R_S$. Consequently, matter can orbit this type of black hole at much smaller radii than in the spherically symmetric case before it reaches the last stable orbit. Therefore, as we will show, much more energy can be extracted when matter falls into a maximally-rotating black hole as compared with the case of a non-rotating black hole.

3.7 The astrophysics of black holes

The properties of black holes are remarkable theoretical results but how do they help us understand the astrophysics of quasars and active galactic nuclei? At first glance, it looks as though we have produced objects which are very good at consuming matter and radiation rather than emitting them. The clue is provided by the binary X-ray sources. The source of energy in these systems is the accretion of matter from the normal primary star onto the compact secondary. This turns out to be an extremely efficient means of generating energy, provided the star is compact enough. All that is necessary is to drop the matter from a very great height onto the surface of a neutron star. A simple calculation shows that, by the time the matter reaches the surface of the neutron star, it acquires a velocity which is a significant fraction of the speed of light. When this matter hits the surface of the star, it is rapidly brought to rest and all the kinetic energy of infall is dissipated as heat. As much as 5% of the rest mass energy of the infalling matter can be converted into heat. This is an enormous amount of energy. It is interesting to compare this rate of energy generation with the amount of energy liberated in the nuclear reactions which power the Sun. In burning hydrogen into helium, 0.7% of the rest mass energy of the hydrogen atoms is converted into energy, whereas in the case of accretion onto neutron stars, about 5% of the rest mass energy of the infalling matter is dissipated as heat – accretion is already about seven times more efficient than nuclear reactions.

In the above example, matter was accreted onto the surface of a neutron star. What happens if instead the matter is accreted onto a black hole? Unlike the surface of a neutron star, a black hole has no solid surface and so how can the energy be released? At first sight, one might think, 'Well, if the matter falls into a black hole, that's not much good – we have simply lost all the matter and not extracted any energy at all.' This is where observations of X-ray binaries provide essential clues, because it is very unlikely indeed that the matter falls directly into the black hole. It is the same problem we encountered in our analysis of the problems of star formation. If the infalling matter acquires even a tiny amount of rotation at a large distance from the black hole, conservation of angular momentum results in the amplification of its rotational energy so that collapse into the black hole is prevented. Just as in the case of star formation discussed in Section 2.5, there must be some means of getting rid of the angular momentum of the matter, if it is to reach the last stable orbit about the black hole.

There is a rather convincing means by which this can happen and that is if the infalling matter forms what is known as an *accretion disc* about the black hole. Such discs are likely to form naturally, because matter can collapse parallel to the rotation axis of the infalling material, just as described in the case of the formation of protostellar discs. Matter can only move into an orbit closer to the black hole by losing rotational energy and the natural way of achieving this is through the viscous dissipation of energy by the material of the accretion disc. This is a beautiful piece of astrophysics and it achieves two desirable things. First, energy is dissipated by viscosity and this frictional energy loss heats up the accretion disc which radiates

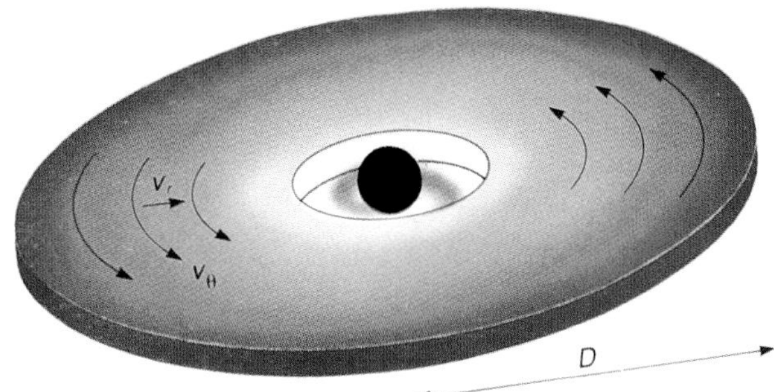

Figure 3.17. Illustrating the structure of a thin accretion disc about a black hole. The material of the disc is in a state of rotation about the black hole and, because of viscosity in the disc, drifts slowly in towards the black hole at velocity v_r. The dissipation of energy associated with the outward transfer of angular momentum in the disc increases towards the central regions. In the case of a spherically symmetric black hole, the last stable orbit about the black hole occurs at $r = 3R_S$, where R_S is the Schwarzschild radius, and the matter collapses irreversibly into the black hole when it passes through this radius.

this energy away from its surface layers. Second, angular momentum is transported outwards by these viscous forces and so the matter gradually drifts in towards the black hole, rotating faster and faster and becoming hotter and hotter as it approaches the last stable orbit (Figure 3.17). This is the means by which the infalling matter can release a large fraction of its rest mass energy as heat before it eventually falls into the black hole.

Let us now quote the theoretical maximum efficiencies for the energies liberated in the accretion of matter onto Schwarzschild and maximally rotating Kerr black holes. In the spherically symmetric case, up to 6% of the rest mass energy of the matter falling into the black hole can be liberated, that is, the efficiency of this process is about ten times greater than that obtained from nuclear processes. If the black hole rotates as fast as possible, a maximally rotating Kerr black hole, about 42% of the rest mass energy of the infalling matter can be extracted. This is an incredible amount of energy – it corresponds to about 50 times more than can be produced by nuclear fusion reactions. Thus, accretion of matter onto black holes is potentially the most efficient source of energy available for powering astronomical objects. There is an additional source of energy associated with the fact that the black hole is rotating. In principle, some of that energy can be tapped to power the activity observed in active galactic nuclei, provided there is some way of linking the rotation of the hole to its surroundings. The theoretical answer is that up to 29% of the rest mass energy of a maximally rotating black hole can, in principle, be made available to the external Universe.

These results are impressive enough but notice that we have obtained much more than simply an extraordinarily powerful source of energy. We have found an energy source which can liberate energy over the *shortest possible time-scale* for an object of a given mass. This result follows from the fact that most of the energy is generated from roughly the last stable orbit

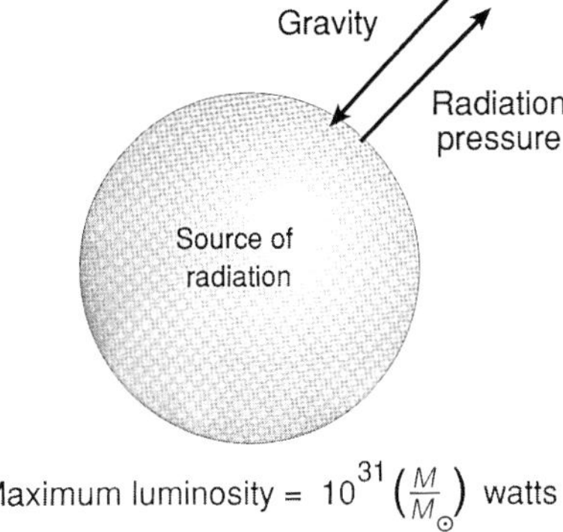

Figure 3.18. Illustrating the origin of the Eddington limiting luminosity.

about the black hole and therefore there could be variations in the intensity of the radiation from the inner regions of the accretion disc on time-scales which are roughly the light travel time across a few times the Schwarzschild radius. Since the Schwarzschild radius R_S is the shortest meaningful physical scale which can be associated with the mass M, we can use the causality argument developed in Section 3.2 to infer that R_S/c must be roughly the shortest time-scale over which the energy generated by a black hole can vary. We therefore obtain two things simultaneously from black holes – enormously powerful sources of energy, in the sense that a significant fraction of the rest mass energy of the matter falling into the black hole can be released to power the active galactic nucleus, and the liberation of this energy over the shortest possible time-scale for an object of that mass.

Before returning to the real world of active galactic nuclei, one more important piece of astrophysics is needed. Radiation transfers momentum as well as energy and so the radiation escaping from any astronomical object exerts an outward pressure. If the luminosity of the source of radiation becomes too great, the radiation pressure associated with it blows away the surrounding material, in particular, the matter which is fuelling the neutron star or black hole (Figure 3.18). This famous calculation was first performed by Arthur Eddington, who showed that there is a maximum luminosity for any source of radiation before it blows away the surrounding matter. Eddington showed that this maximum luminosity depends only upon the mass of the object. Specifically, the expression for the *Eddington limiting luminosity* is

$$L_{\mathrm{Edd}} = 10^{31}\left(\frac{M}{M_\odot}\right)\mathrm{W}.$$

For comparison, the luminosity of our own Sun is 3.9×10^{26} W and so it radiates well below the limiting luminosity for solar mass stars. On the other hand, the X-ray luminosities of the Galactic X-ray sources extend up to the Eddington limiting luminosity, assuming that the masses of the parent bodies of the X-ray sources are between about 1 and 10 $M_\odot$. This is very good evidence that, in the case of the X-ray sources at least, there exist objects which are powered by accretion and which can radiate at more or less the maximum luminosity permitted by theory.

3.8 Observational evidence for black holes

What is the observational evidence for the existence of black holes in astronomical systems? The problem is that isolated black holes are very difficult to detect, except through their gravitational influence upon their surroundings. The best approach for finding stellar mass black holes is to search for them in binary X-ray sources containing invisible companions, which are inferred to have masses greater than 3 $M_\odot$. Such invisible companions can only be black holes. A great deal of effort has been devoted to analysing whether or not any of the X-ray binary sources contain black holes. Four X-ray binaries have been discovered in which there is reasonably convincing evidence for the presence of black holes. The characteristics of these systems are that they are high luminosity X-ray sources, which must be radiating close to the Eddington limit for stellar mass objects, and which do not contain X-ray pulsars. They exhibit rapid variability or 'flickering' of their X-ray intensities which would be consistent with the X-rays originating from close to a black hole since the minimum time-scale for such variability is roughly $t \sim R_S/c \sim 0.01\ (M/M_\odot)$ milliseconds. Evidence of flickering on the time-scale of milliseconds has been observed in a number of these sources.

The masses of the unseen companions are derived by the classical astrometric techniques of the analysis of binary star orbits. It is probable that the masses of the unseen companions in the four binary X-ray sources illustrated in Figure 3.19 are greater than 3 $M_\odot$ and so they cannot be white dwarfs or neutron stars. The estimated masses of the binary components are model dependent and, in particular, depend upon knowledge of the inclination of the binary orbit to the plane of the sky and the mass of the visible companion. The masses of the primary stars and black holes shown in Figure 3.19 are best estimates but there is some uncertainty about the precise values. This is a very important and exciting area of astrophysics and confirmation is urgently required that there are indeed black holes in these binary systems. These sources provide us with a unique opportunity for studying the behaviour of matter in very strong gravitational fields.

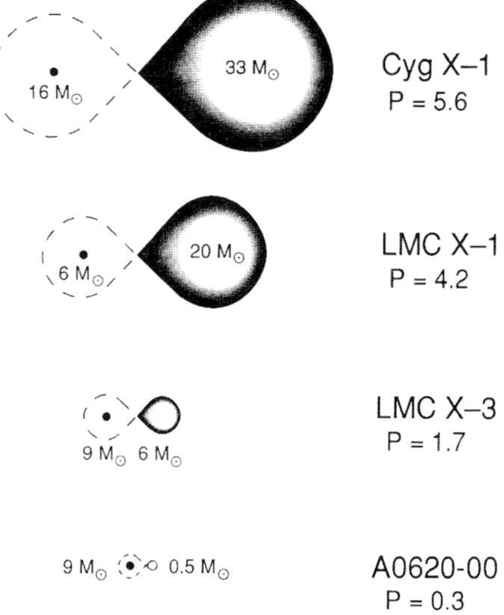

Figure 3.19. Schematic sketches, to scale, of plausible models for the four X-ray binary systems, in which the invisible companion is probably a stellar mass black hole.

The other objects, in which a convincing case for the presence of black holes can be made, are the active galactic nuclei. The analyses carried out in Sections 3.6 and 3.7 show that many of the properties of the emission from the vicinity of black holes depend upon their masses. Let us study two beautiful examples which illustrate how the masses of active galactic nuclei can be measured. The first example concerns the galaxy NGC 4151 (Figures 1.12 and 1.13). This galaxy contains one of the most active nearby nuclei and there are rather dense gas clouds close to the nucleus, which are identified by the strong broad emission lines in its spectrum. The properties of the active nucleus and surrounding clouds have been studied in a beautiful set of observations made by the ultraviolet space telescope known as the International Ultraviolet Explorer (IUE). Convincing evidence was found that the gas clouds are heated and excited by the ultraviolet radiation emitted by the active nucleus. The remarkable feature of this set of observations was that it was possible to measure the time delay between a burst of ultraviolet radiation occurring in the nucleus and the emission of intense line radiation by the surrounding clouds. What is happening physically is that the ultraviolet radiation from the nucleus ionises and excites the atoms in the surrounding clouds and the emission lines are due to the recombination of the excited ions. These observations enabled an estimate of the distance of the clouds from the nucleus to be made, since the radiation travels this distance at the speed of light. The velocities of the clouds about the nucleus were estimated from the Doppler shifts of the broad emission lines. Combining these observations, the mass within the orbits of the gas clouds can be estimated. Exactly the same type of calculation enables the mass of the Sun to be measured, if the Earth's distance from the Sun and its orbital velocity, v, are known. As in that calculation, balancing the centrifugal force acting on a cloud of mass m at distance R from the nucleus by the gravitational attraction of the nucleus, of mass M, we find that

$$\frac{mv^2}{R} = \frac{GMm}{R^2}$$

and hence the mass of the nucleus can be found:

$$M = \frac{v^2 R}{G}.$$

Using these observations, it was found that there must be a compact object with mass at least $10^8 M_\odot$ in the nucleus of the galaxy NGC 4151. Similar observations have now been made for a number of other active galactic nuclei – the nuclei have masses which are about $10^6 - 10^9 M_\odot$.

Another beautiful example concerns recent observations of the nearby active galaxy M87 by the Hubble Space Telescope. Figure 3.20(*a*) is a picture of the central region of M87, which is one of the brightest giant elliptical galaxies in the nearby Virgo cluster of galaxies. The intense optical nucleus can be seen to the bottom left of the main image and extending to the top right is the famous jet, which was first noted by Heber Curtis in 1918. The important feature of the Hubble Space Telescope image is the presence of gas clouds in the central regions of the galaxy, despite the fact that it is classified as an elliptical galaxy. In particular, there is gas, which emits strong emission lines, close to the nucleus in the form of a disc and the velocities of this gas on either side of the nucleus have been measured (Figure 3.20(*b*)). These spectroscopic observations have shown that the gas is approaching us on one side of the disc and moving away from us on the other. It is inferred that the disc is rotating about the nucleus of M87 at a velocity of about 550 km s^{-1} and that the nucleus must have mass about $3 \times 10^9 M_\odot$. This is a far greater mass than can be attributed to stars in the central core of the galaxy and it is natural to attribute it to the presence of a supermassive black hole.

We have derived three important relations concerning the properties of black holes as the energy sources for active galactic nuclei and these provide strong constraints upon models. These results are:

- Accretion is a very efficient means of energy production with efficiencies amounting to 5 to 40% of the rest mass energy of the infalling material.
- The shortest time-scale of variability for an object of mass M is given roughly by
$$t = R_S/c = 10^{-5}\left(\frac{M}{M_\odot}\right)\text{s}.$$
- The maximum luminosity of the source is limited by radiation pressure according to the Eddington limit
$$L < 10^{31}\left(\frac{M}{M_\odot}\right)\text{W}.$$

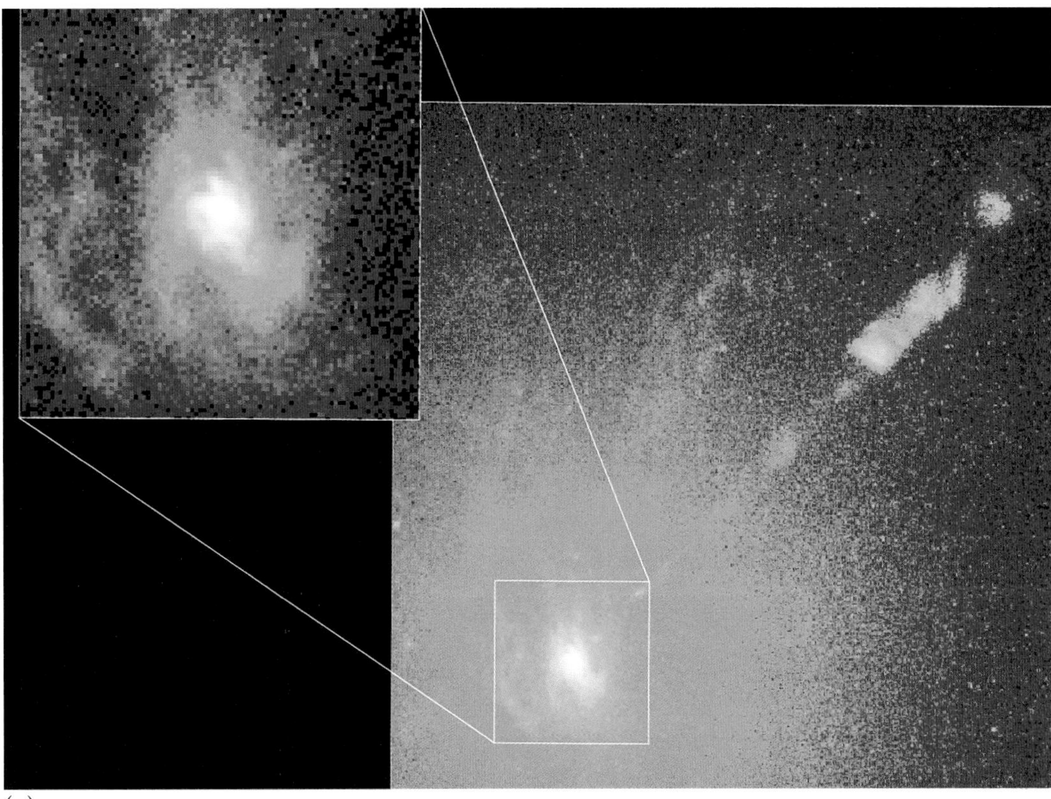

(a)

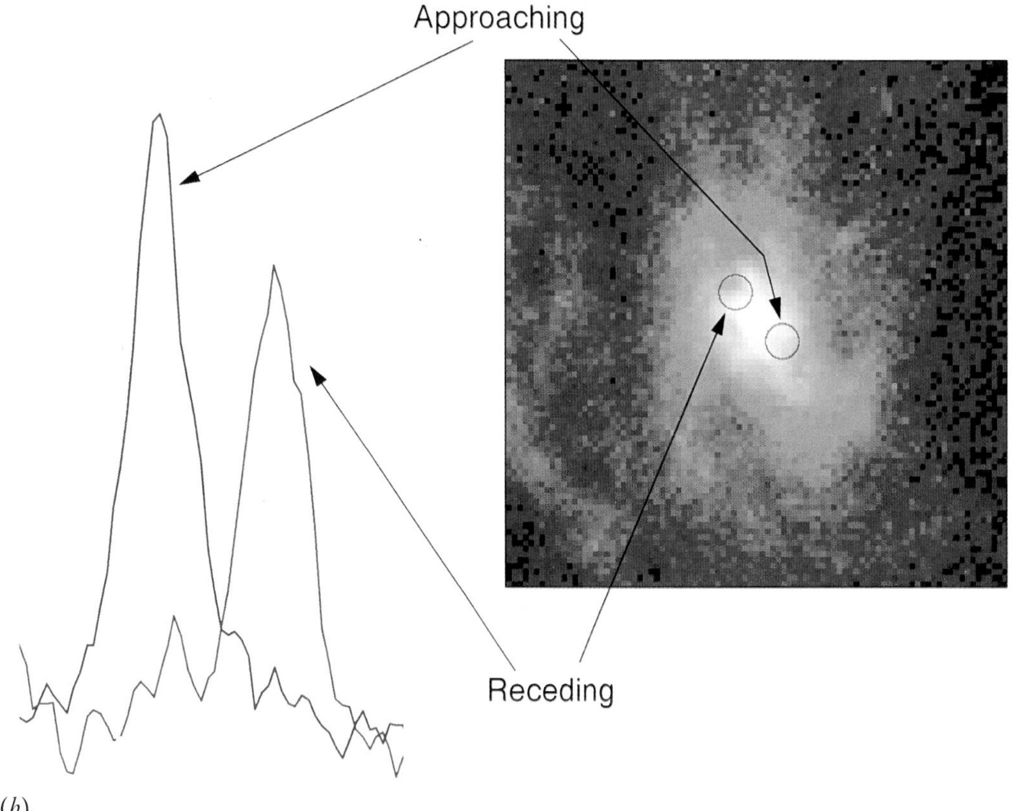

(b)

Figure 3.20. (*a*) Images of the central region of the active galaxy M87 taken by the Wide Field Planetary Camera 2 of the Hubble Space Telescope. The large image shows the nuclear regions as well as the prominent optical jet. The inset shows the nucleus and central gaseous disc of the galaxy in more detail. (*b*) The spectra of emission line gas in the central disc of the galaxy on either side of the nucleus at the locations indicated in the diagram. The gas on one side of the galaxy is strongly redshifted and that on the other side blueshifted, showing that the disc is rotating at a speed of about 550 km s^{-1}.

Do active galactic nuclei satisfy these criteria? The answer is displayed in Figure 3.21 which shows mass estimates for a range of active galactic nuclei derived from both dynamical arguments and from the variability of the emission from the nuclei (Figure 3.21(*a*)). It can be seen that these mass estimates are in reasonable agreement. Then, using these masses, each active nucleus can be plotted on a mass–luminosity diagram to find out whether or not it approaches the region of the diagram which is forbidden according to the Eddington criterion. It can be seen from Figure 3.21(*b*) that the active galactic nuclei do not fall in the forbidden region. These diagrams show that even the most extreme quasars can be accommodated within the simplest black hole models for active galactic nuclei.

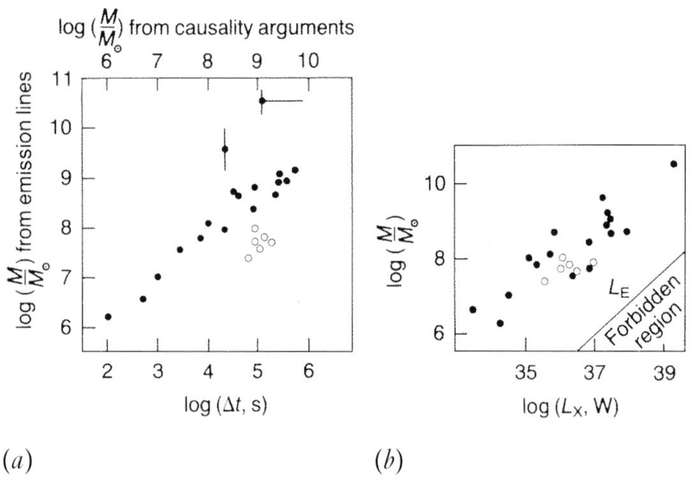

Figure 3.21. (*a*) Comparison of mass estimates for active galactic nuclei from the variability of their X-ray emission and from dynamical estimates due to Wandel and Mushotzky. The solid dots are quasars and type 1 Seyfert galaxies. The open circles are type 2 Seyfert galaxies. (*b*) Comparison of the inferred masses and luminosities of active galactic nuclei with the Eddington limiting luminosity. It can be seen that all the points lie well away from the Eddington limit.

3.9 The big problems

In my view, the above analyses are persuasive arguments in favour of the proposition that supermassive black holes must be present in active galactic nuclei. It will have been noted, however, that the arguments refer to very general properties of active galactic nuclei rather than to the detailed properties of particular classes of object. Indeed, this is the strength of this line of reasoning. Can we use the picture of supermassive black holes to account for the many diverse phenomena observed in active galaxies? The honest answer is, 'Not really!' It is a much more complex problem to account in detail for all the high energy astrophysical phenomena observed in active galactic nuclei. Let us summarise some of the relevant information which the different wavebands provide about the essential ingredients of any successful model.

- Besides the key information which optical, infrared and ultraviolet observations provide about the variability of the radiation from active galaxies, there is a great deal of spectroscopic information about both the strong emission lines from gas clouds in the vicinity of the nucleus and the underlying continuum radiation which extends all the way from the ultraviolet to the infrared waveband. The continuum radiation does not have the spectrum of thermal radiation but is probably the synchrotron radiation of high energy electrons radiating in the strong magnetic fields in active galactic nuclei. The evidence of active galaxies such as NGC 4151, which was described in Section 3.8, indicates that the intense ultraviolet continuum radiation from the nucleus is responsible for ionising and exciting the surrounding gas clouds. Analyses of the spectra of active galactic nuclei show that there are clouds of a wide range of different densities in the vicinity of the nucleus. Closest to the nucleus are dense gas clouds which are the source of the broad variable emission lines seen in galactic nuclei such as NGC 4151. Further out are more diffuse clouds moving at lower velocities about the nucleus. The origin of these gas clouds is not understood but it is evident that some source of gas is needed to power the active galactic nucleus.

Another important result provided by optical observations concerns the most violently variable active galactic nuclei, which are associated with the sources known as BL-Lac objects or blazars. These objects vary in intensity on the time-scale of a day or less and yet some of them are as distant as the quasars. One of their distinctive features is that their optical spectra are featureless, consisting solely of smooth continuum radiation. The radiation is also often strongly polarised.

- The radio observations show a remarkable range of activity associated with the energetic jets which power the extended outer radio components. The example of Cygnus A (Figure 3.3) is of special interest because it can be seen that the outer radio components are supplied with energy by very long, narrow jets of radio emitting material. These jets are essential to supply energy to the compact 'hot-spots' observed towards the outer edges of the radio components. The radio jets can be traced back to the nuclear regions of the parent galaxy. Very long baseline interferometric techniques have been used to study the structure of the jet in Cygnus A on the scale of 0.001 arcseconds and, even on that very fine scale, the jet is still pointing in the direction of the long narrow jet and the outer hot-spots. This type of morphology is found in many of the most powerful radio galaxies and radio quasars. Thus, an essential ingredient of the nuclear regions is some means of producing a highly collimated jet of radio emitting material which ultimately powers the outer radio lobes.

As if there were not sufficient problems already, one of the most amazing results of modern astrophysics has been the discovery that, on a very fine scale, some of the radio jets observed in the very centre of compact radio quasars seem to be expelled at velocities exceeding the speed of light. In Figure 3.22(*a*), radio maps of the central regions of the radio quasar 3C 273 are shown and it can be seen that the radio components seem to have separated from each other by 25 light years in only three years, corresponding to an apparent speed of separation of over eight times the speed of light. Such velocities are referred to as *superluminal velocities*. They are common phenomena among the compact luminous radio sources. According to Einstein's special theory of relativity, such physical velocities are not allowed. Almost certainly what is happening is that the jets of radio emitting material are moving at a very high but subluminal velocity at a small angle to the line of sight. How this can explain the superluminal velocities is illustrated in Figure 3.22(*b*). The diagram is drawn to scale and the jet of material is ejected at a speed of 0.98 times the speed of light at an angle of 11.5° to the line of sight. As illustrated in the diagram, the component is then observed to separate from the stationary nucleus at five times the speed of light. Thus, the apparent sideways motion of the jet can exceed the speed of light, without violating any of the rules of the special theory of relativity. The consequence is, however, that the jets must escape from the nucleus at very high speeds indeed. This is very good news, since it demonstrates that these radio jets can transport material at very high speeds from the nucleus and these can power the extended radio lobes seen in powerful radio sources.

- The X-ray and γ-ray intensities of active galactic nuclei are also variable and this information was used above to determine the dimensions of the emission regions. The most recent example of the importance of relativistic motion in active galactic nuclei has come from observations made with the Compton Gamma-Ray Observatory. The high energy γ-ray telescope has detected variable γ-ray emission from a number of the most luminous, compact radio quasars, in particular, from those which exhibit evidence for superluminal velocities. This is an intriguing result because, if the radiation were not beamed, the energy density of the γ-ray emission would be so great in the source region that all the γ-rays would be degraded by creating electron–positron pairs, before they could escape from the source region. If the material responsible for the γ-ray emission moves out from the nucleus at the same relativistic velocities as the radio jets, this problem can be resolved. It seems that relativistic beaming is an essential ingredient of the most active galactic nuclei.

It is evident that some of the above features of active galactic nuclei may be thought of as primary ingredients and others as a consequence of the interaction of the primary ingredients with the environment of the nucleus and the host galaxy. Thus, it seems essential that the continuum infrared, optical and ultraviolet radiation originate close to the active nucleus itself. Likewise, the material responsible for the jets emanating from the nucleus must be primary ingredients. The interaction of these components with clouds in the vicinity of the nucleus and with the ambient interstellar and intergalactic medium are the origin of some of the phenomena described above. In Figure 3.23, I have put together a sketch of what the nuclear regions of a quasar or active galactic nucleus may look like, but it is highly schematic. Simple thin accretion discs were illustrated in Figure 3.17 and these can account for the properties of certain types of galactic source, for example, X-ray binaries and various types of less extreme forms of binary star systems, such as the cataclysmic variables. The problem in the case of active galactic nuclei is that the luminosities are very great indeed and so the rate at which matter is accreted must be much greater than in the case of galactic sources. The radiation emitted from the inner regions of the accretion disc is so intense that its radiation pressure inflates the inner regions into what is known as a thick accretion disc. A sketch of one of these is shown schematically in Figure 3.23 and it has 'funnels' along the rotation axis which might be responsible for collimating the jets of material from the nucleus. Unfortunately, stability analyses which have been made of thick accretion discs have suggested that these are unstable configurations.

It is clear that some of the observed properties of active galactic nuclei, such as superluminal velocities of ejection, depend upon the angle at which the source is observed relative to the line of sight. This concept raises the intriguing question of to what extent orientation effects influence the observed appearance of active galaxies in general. A good case can be made that the powerful radio galaxies and radio quasars may be the same type of object, observed at different angles to the line of sight. The same seems to be true of the distinction between the active galaxies known as Seyfert I and Seyfert II galaxies. The idea is that there is an obscuring torus about the nuclear regions of the galaxy which prevents the intense light

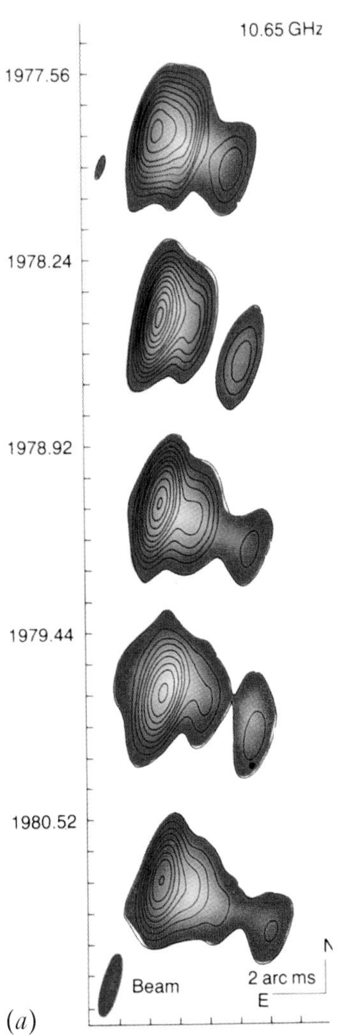

(a)

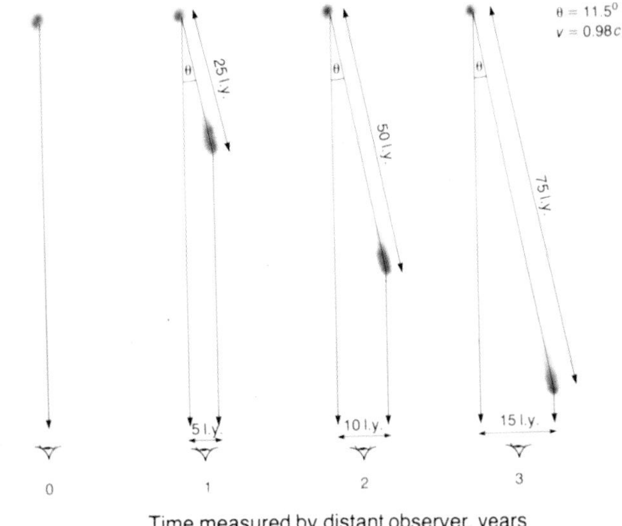

(b)

Figure 3.22. (a) Very long baseline interferometry images of the nucleus of the radio quasar 3C 273 over the period 1977 to 1980. The radio component on the right appears to travel a distance of 25 light years in a period of only three years, implying a superluminal velocity of about eight times the speed of light. (b) A diagram to scale illustrating the appearance of a radio source component ejected at 0.98 of the speed of light at an angle of 11.5° to the line of sight, as observed from above. The apparent superluminal motion of the component is associated with the fact that it almost catches up with the radiation emitted at earlier times.

from the nucleus being observed, if the axis of the torus lies at a large angle to the line of sight. On the other hand, if we observe the nucleus at a small angle to the axis of the torus, an unobscured quasar in the nucleus of the galaxy is seen. It is then natural to assume that the jets of material responsible for powering the radio sources are emitted along the axis of the system. These jets move at speeds close to the speed of light along the axis of the source and are responsible for the superluminal radio emission and for the intense γ-ray emission. The inference that these emissions are strongly beamed, often termed *relativistic beaming*, has important consequences for the interpretation of some of their more extreme properties. For example, it is quite plausible that the BL-Lac objects have their extreme properties because the radio and optical emitting material is observed along the direction of ejection of a relativistic jet. A schematic model, which shows how these features can be combined into a unified model for radio galaxies, quasars, BL-Lacs (blazars), Seyfert I and Seyfert II galaxies is illustrated in Figure 3.24. It remains to be seen how well this picture can explain the many diverse aspects of active galactic nuclei.

What is the origin of the supermassive black holes in the nuclei of active galaxies? It is probably inevitable that they form by one means or another. The number density of stars increases dramatically towards the centre of a galaxy and the central region may be thought of as of a compact star cluster. In such a system, gravitational encounters between the stars cause the cluster to lose its most energetic members and the cluster shrinks until eventually the stars come so close together that they collide and coalesce to form massive systems which collapse to form black holes. Any gas liberated in a galaxy tends to fall towards the centre of the galaxy. Thus, provided the matter can lose its angular momentum, it will accumulate at the very centre of the galaxy. This matter can be accreted onto any pre-existing black hole, thus increasing its mass. As the mass of the hole increases, it can begin to disrupt stars passing too close to it, consuming some of their mass and so increasing the mass of the hole. Eventually, when the black hole becomes massive enough, it can swallow stars whole without disrupting them. Thus, there are convincing reasons why supermassive black holes might form in the nuclei of galaxies but which of these is the most important process is not at all clear. What is certain is that quasars are among the most distant objects we know of in the Universe, the

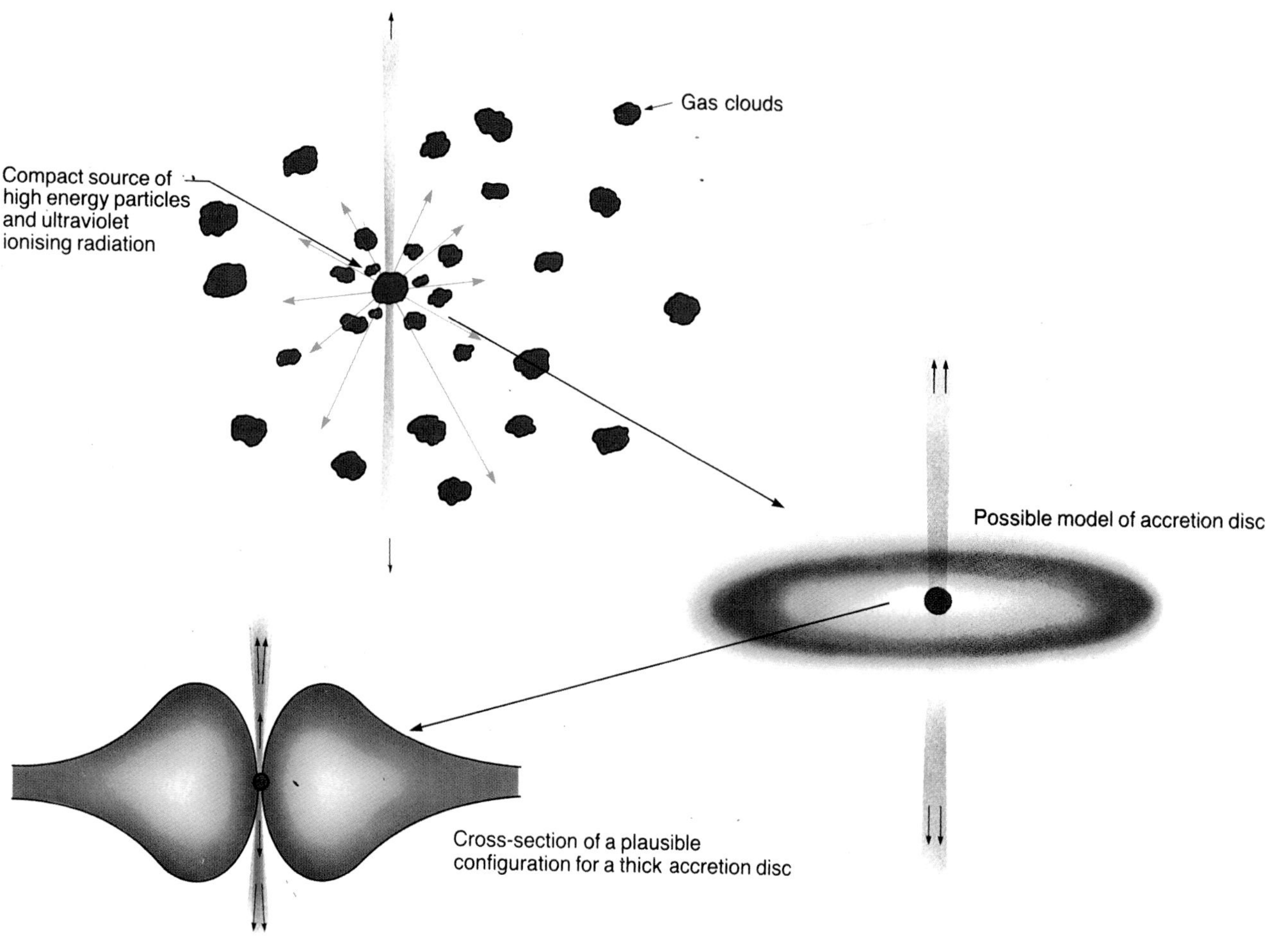

Figure 3.23. A schematic model showing the necessary ingredients of a model of a quasar or active galactic nucleus. There must be a compact source of radiation and high energy particles in the nucleus. The nucleus is surrounded by gas clouds which are heated and excited by the ultraviolet radiation from the nucleus. The inserts show a possible structure for the accretion disc and the thick disc surrounding the black hole. These may be responsible for collimating the beams of high energy particles. It is not at all certain that this model of a thick disc is stable.

most distant of them emitting their light when the Universe was less than one fifth of its present age. Therefore, the formation of supermassive black holes must have been possible when the galaxies were very much younger than they are today.

So, what progress have we made? It seems highly probable that there are supermassive black holes in active galactic nuclei – black holes have exactly the right properties to account for their extreme luminosities and their short time-scales of variability. We therefore have the opportunity of studying the behaviour of matter in strong gravitational fields which cannot be produced in the laboratory. But there are no easy solutions to the many other facets of active galactic nuclei. The successful theory will have to explain the origin of relativistic jets, as well as the origin of the material of accretion discs and obscuring tori. After more than 30 years, the more active galactic nuclei have been studied, the more complex the problems seem to have become. Perhaps it is not so surprising that understanding their properties is not a trivial business. They are buried in the centres of galaxies where we know the environment is complex and it is not perhaps surprising that the supermassive black holes interact in somewhat unexpected ways with their surroundings. There are many problems which still need to be studied both observationally and theoretically. This is very good news since it means that there is plenty of research to be carried out by future generations of astronomers.

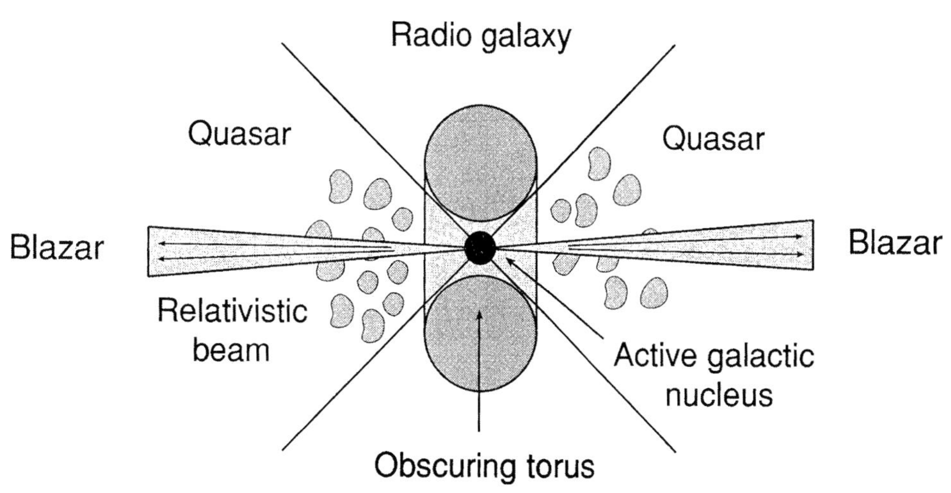

(*a*)

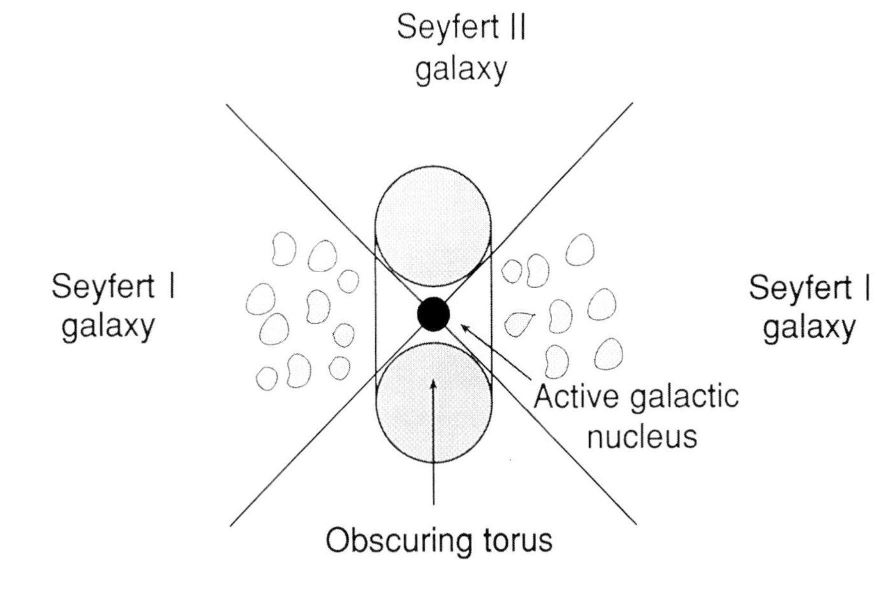

(*b*)

Figure 3.24 Schematic models showing how different types of active galaxies may be accommodated within a unified scheme in which the differences between them are attributed to observing them at different angles to the line of sight. It is believed that the radio galaxies and radio quasars are associated with giant elliptical galaxies, while the Seyfert galaxies are principally spiral galaxies. (*a*) A unified picture for radio galaxies, quasars and blazars. (*b*) A unified picture for Seyfert I and Seyfert II galaxies. In the Seyfert I galaxies, broad and narrow emission line regions are observed, the broad line regions originating close to the nucleus itself. In Seyfert II galaxies only the narrow line regions are observed.

4 The origin of galaxies

4.1 The evolution of galaxies oversimplified

The problems of the origin and evolution of galaxies are quite different from those involved in the formation and evolution of the stars, which were discussed in Chapter 2. In that chapter, the life cycle of stars was described – their birth in dense dusty regions of interstellar space, their long lifetimes as main sequence stars and their violent deaths – what I have called the great cosmic cycle of the birth, life and death of the stars. We have therefore gained some understanding of the internal workings of the constituents of galaxies. There is a continual interchange of matter between the stars and the interstellar gas. Stars form in dense regions of interstellar gas and, when they die, they return processed material to replenish the interstellar medium. As a result, there is a great deal of rich structure in the interstellar medium at a wide range of different temperatures, from very cool gas in the dusty regions to very hot gas, heated up by the explosions of stars. Large scale gravitational perturbations as well as dynamical processes, such as supernova explosions, can result in the formation and cooling of giant molecular clouds within which new generations of stars are formed. As a result, the stellar content of galaxies depends upon their star formation histories.

Some impression of the effect of different star formation histories upon the appearance of a galaxy can be obtained from the Hertzsprung–Russell diagrams of star clusters of different ages, which we introduced in Section 2.3. The H–R diagram of the old globular cluster 47 Tucanae was shown in Figure 2.11(*b*). In this cluster, star formation ceased a very long time ago, soon after the cluster was formed and all stars with masses greater than roughly the mass of the Sun, which has a lifetime of about 10 billion years, have evolved off the main sequence. Stars with mass roughly the mass of the Sun are currently evolving off the main sequence and these are the stars which make up the giant branch, which is beautifully defined in Figure 2.11(*b*). The total light of the cluster is dominated by the luminous stars on the red giant branch and so the cluster as a whole has a red appearance.

The H–R diagrams for younger star clusters were illustrated in Figure 2.10. The main sequence termination point occurs progressively further and further up the main sequence with decreasing age because, in younger clusters, stars more massive than the Sun are still completing their evolution on the main sequence. The light of the clusters is dominated more and more by massive stars on the main sequence and these dominate the light of the cluster. Thus, the younger the cluster, the bluer its colour as massive young main sequence stars make a greater contribution to the total light.

The colours of galaxies can similarly be broadly understood on the basis of their different star formation histories. In the elliptical galaxies, there is little ongoing star formation and their light is dominated by old stars, similar to those responsible for the colours of old globular clusters, such as 47 Tucanae. In contrast, in spiral galaxies, there is ongoing star formation within their spiral arms and so their colours are much bluer than those of elliptical galaxies. The bluest regions are those in which the youngest stellar populations are found. True colour images of typical elliptical and spiral galaxies are illustrated in Figure 4.1, showing the different colours of the old spheroidal stars and the blue spiral arm stars. The overall

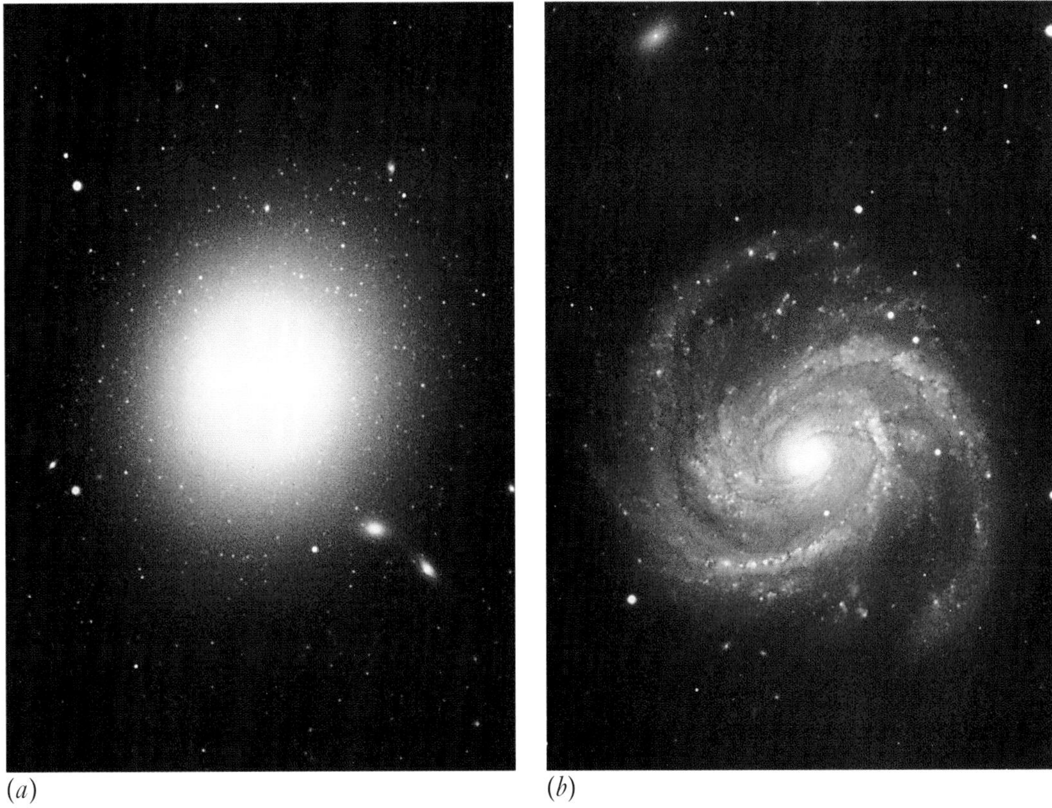

Figure 4.1. True colour images of: (a) the giant elliptical galaxy M87 (NGC 4486) in the nearby Virgo cluster of galaxies. The galaxy is surrounded by a cloud of globular clusters; (b) the spiral galaxy M100 (NGC 4321), showing the differences in colour between the old red stellar populations of the spheroidal components of both types of galaxy and the young blue populations in the spiral arm regions of spiral galaxies.

colours of spiral galaxies can be accounted for, if it is assumed that star formation has continued at a more or less constant rate throughout the bulk of their lifetimes.

A major growth area in the astrophysics of galaxies has been the construction of theoretical models for the evolution of the stellar content of galaxies of different types from the time of their birth to the present day. The evolutionary histories of stars of all masses need to be known and then the evolution of the stellar content of the galaxy as a whole can be worked out on the basis of different assumptions about the star formation history of the galaxy. These studies are very important for understanding the properties, not only of nearby galaxies but also of very distant galaxies which are observed at earlier stages in their history and which can now be observed in surveys of very faint, and hence very distant, galaxies.

In these models of galactic evolution, it is assumed that the galaxies are closed systems, meaning that the stars in the galaxy are formed as part of the great cosmic cycle described above without any new input of stars or gas from outside the galaxy – we now know, however, that this is unlikely to be the case. Many galaxies show clear evidence of gravitational and dynamical interactions with nearby galaxies and with the gas between the galaxies, what is known as the *intergalactic medium*. There is now good evidence that interactions with other galaxies can lead to enhanced rates of star formation. Collisions between interstellar gas clouds when galaxies collide result in the formation of dense regions of dust and gas and consequently to enhanced rates of formation of new stars. This process is most beautifully illustrated by observations of colliding and interacting galaxies made by the IRAS satellite. These show that the most luminous galaxies in the far infrared region of the spectrum are very often

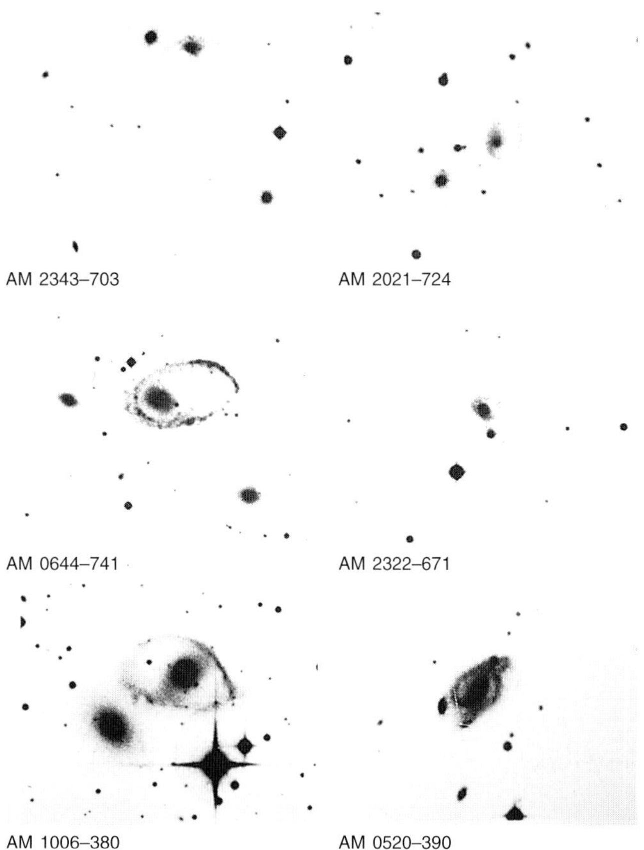

Figure 4.2. Examples of interacting or colliding galaxies from the *Catalogue of Southern Peculiar Galaxies and Associations* prepared by Halton Arp and Barry Madore.

colliding or strongly interacting galaxies. Since intense far infrared emission is the signature of active star formation, by the processes described in Sections 2.4 and 2.5, these observations indicate that collisions between galaxies can initiate major bursts of star formation in galaxies. For these reasons, these luminous IRAS galaxies are known as *starburst galaxies*.

Although most galaxies can be classified as spiral, elliptical or lenticular galaxies, there are a few weird galaxies which have very strange appearances indeed. Some examples of these 'interacting galaxies' are illustrated in Figure 4.2, which is taken from the catalogue of interacting galaxies in the southern hemisphere prepared by Halton Arp and Barry Madore. It is certain that gravitational encounters between galaxies are responsible for some of these remarkable structures. Let us look at one of the most famous examples of these interacting galaxies, the system known as the Antennae (Figure 4.3(*a*)). The system seems to consist of the collision of two galaxies but the amazing features of the picture are the two huge 'antennae' which have been produced in the collision. At first sight, it is difficult to imagine how such a strange structure could be produced by the collision of two galaxies, but it can be done, as has been beautifully demonstrated by the computer simulations of Alar and Yuri Toomre. In their simulations, they have represented the two colliding spiral galaxies by rotating discs of stars and then followed in a computer how the stars in each of the discs behave under the mutual gravitational forces acting between all the stars in the galaxies. By studying many simulations of close encounters between these model galaxies, they found that, if the galaxies passed by each other in the same sense as the rotation of each disc of stars, that is, in the 'prograde'

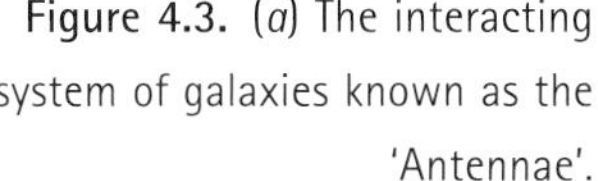

Figure 4.3. (*a*) The interacting system of galaxies known as the 'Antennae'.

direction, the outer rings of stars are ripped off each of the galaxies, resulting in the formation of spectacular 'tails', similar to those seen in the Antennae (Figure 4.3(*b*)). The reason for these remarkable structures is that, in a prograde encounter, the stars in the outer rings feel the same outward force acting in the same direction for a prolonged period.

The systems illustrated in Figures 4.2 and 4.3 are spectacular examples of collisions or, more precisely, gravitational interactions between galaxies. Similar encounters are likely to be quite common on a less spectacular scale in many galaxies. This is because we now know that isolated galaxies are very rare indeed and that most galaxies are members of small groups or clusters of galaxies. Therefore, the probability of encounters between galaxies is much increased as compared with what would be expected if the galaxies were randomly distributed in the Universe. Even within the Local Group of galaxies, of which our Galaxy and the Andromeda Nebula are the dominant members, the Large and Small Magellanic Clouds are currently being torn apart by the gravitational influence of our Galaxy and eventually some of this debris will be assimilated by our Galaxy. At the other extreme, in rich clusters of galaxies such as the Pavo cluster (Figure 1.15), a supergiant elliptical galaxy has formed at its centre. It is likely that collisions between galaxies have played an important role in creating its massive distended form. Gravitational encounters between galaxies result in the more massive members of the cluster losing energy and so sinking towards the centre of the cluster. There, they interact and coalesce creating the typical supergiant galaxies commonly observed

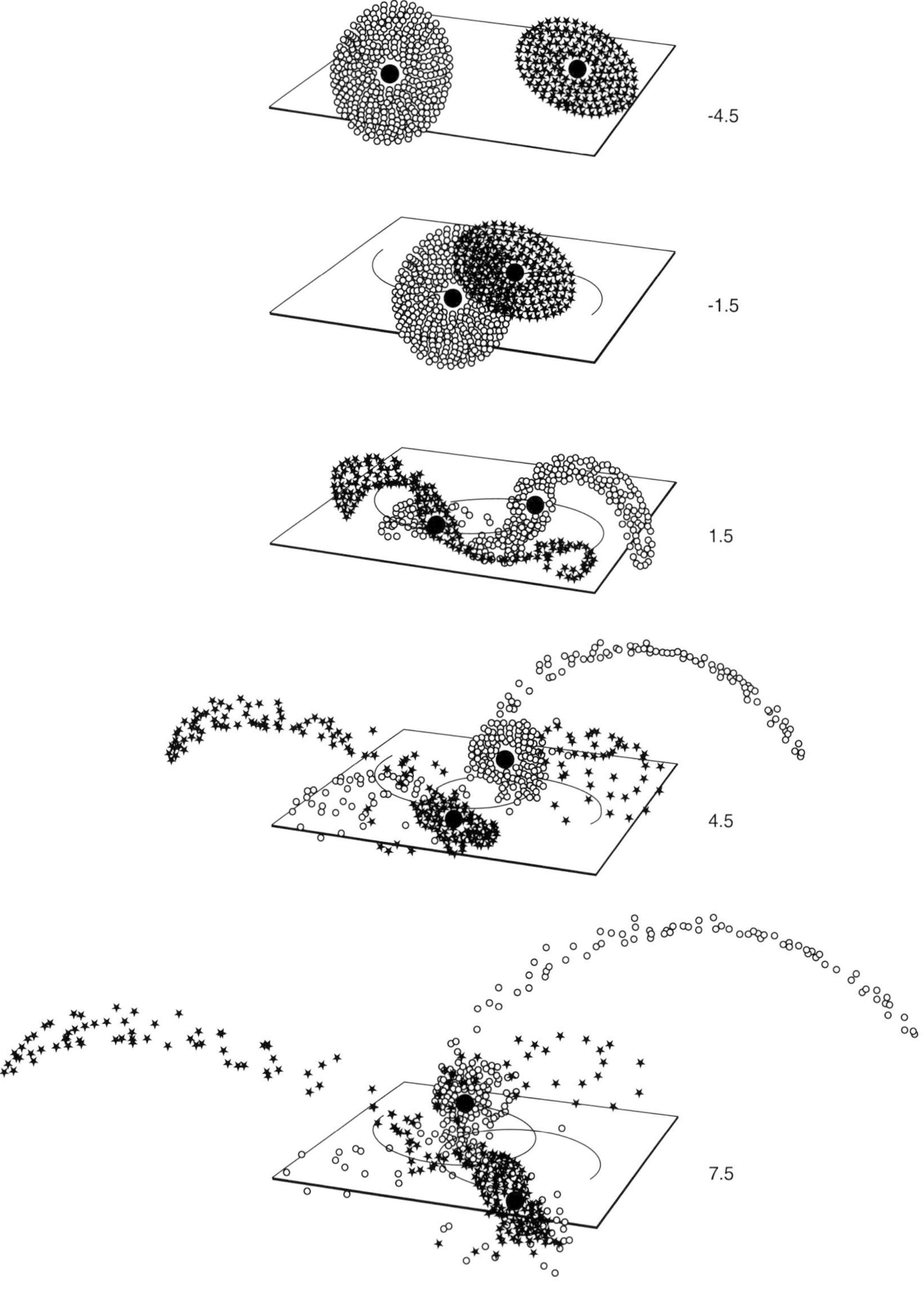

Figure 4.3. (*b*) Computer simulations of a collision between two spiral galaxies carried out by Alar and Yuri Toomre. The encounter takes place in a 'prograde' sense so that the paths of the galaxies are in the same sense as the rotation of the discs of stars. The outer rings of stars are torn from each galaxy resulting in the formation of long 'tails' of stars.

in the richest clusters. Furthermore, if galaxies away from the centre collide, stars and gas clouds may be ripped off, as in the case of the Antennae, and some of this material will sink towards the centre of the cluster where it will contribute to increasing the mass of the central galaxy.

The understanding of how galaxies and clusters of galaxies evolve has advanced dramatically over recent years but there remain many problems to be solved. For example, we do not know the rate at which the heavy elements were built up in galaxies and how this process differs from place to place within a galaxy. How important is the infall of matter from intergalactic space and how significant a role do gravitational encounters play in the evolution of spiral and elliptical galaxies? How are the spiral arms observed in normal and barred spiral galaxies maintained? These problems of the astrophysics of galaxies are far from being solved and they are all currently the subject of intensive research.

The really big problem, which is the subject of this chapter, is, 'How did it all get started in the first place?', in other words, how was the gas out of which the galaxies formed brought together so that the processes of star formation and the great cosmic cycle could get underway? It can be assumed that, once the stars are formed, the galaxies evolve as described above to produce the diversity of structures observed within galaxies. At first sight, our goal does not appear to be particularly ambitious. We recall the remarkable efficiency with which the Jeans instability, which was discussed in Chapter 2, enables infinitesimally small perturbations in the gas density of a cloud to grow to high densities in a finite time under the attractive influence of gravity. It seems as if all we have to do is to find some means of creating infinitesimally small perturbations on the scales of galaxies and clusters of galaxies in the early Universe and then, exactly as in the case of star formation, gravity will do the rest for us. It appears to be a straightforward story but, as we will show, it turns out to be not nearly as simple as it seems. In fact, we will show that it leads to one of the four fundamental problems of modern cosmology.

To appreciate how this problem comes about, we have to understand the background against which small inhomogeneities in the matter distribution in the Universe have to grow. In particular, it is the fact that the Universe is expanding as a whole which changes the picture dramatically. It is high time we discussed the evidence for the expansion of the distribution of galaxies, one of the great discoveries of 20th century astronomy.

4.2 Hubble's law and the expansion of the Universe

I always find it remarkable that our knowledge that the galaxies are extragalactic systems is no more than 70 years old. Astronomers and philosophers of the 18th and 19th centuries had speculated that the spiral nebulae were distant 'island universes', similar to our Milky Way, but there was no observational evidence to support this picture, since their distances could not be measured. It was only in the first two decades of this century that astronomers began the systematic quantitative study of the structure of our Galaxy. A major controversy arose between the years 1915 and 1925, nowadays referred to as the 'Great Debate', which centred on two related questions: first, what is the physical size of the Galaxy and, second, do the spiral nebulae belong to our Galaxy or are they extragalactic systems? The controversy was resolved conclusively by Edwin Hubble in 1925, when he announced a convincing measurement of the distance of our nearest spiral neighbour in space, the Andromeda Nebula or M31 (Figure 1.6), using Cepheid variable stars. These observations showed that M31 is indeed an extragalactic object. This marked the beginning of extragalactic research and observational cosmology. Within a year, Hubble published a comprehensive paper describing the properties of different types of galaxy and made the first estimate of the average mass density of the

Universe in the form of galaxies. He recognised the cosmological significance of this measurement and compared it with the expectations of Einstein's static model of the Universe, which was published in 1917. The significance of Einstein's paper was that it was the first fully self-consistent model of the large scale structure of the Universe and the basis of the model was his general theory of relativity, which had only been completed in 1915. Hubble's subsequent career was devoted to the study of galaxies and their large scale distribution in the Universe.

In 1929, Hubble discovered the famous law which bears his name, *Hubble's law*. From studies of a sample of only 24 nearby galaxies at distances less than 6 million light years from our own Galaxy, he found that, the more distant a galaxy, the greater its velocity away from our own Galaxy – these velocities, measured by the Doppler shifts of spectral lines towards the red end of the spectrum, are called the *recessional velocities* of galaxies. Hubble's data suggested that the recessional velocity of a galaxy is proportional to its distance. Within a few years, Hubble and Milton Humason established that the same relation held good to very much greater distances. A modern version of these pioneering observations is shown in Figure 4.4, which is Allan Sandage's famous redshift–magnitude relation for the brightest galaxies in rich clusters of galaxies. Along the horizontal axis, the apparent magnitudes of the galaxies are plotted. *Apparent magnitude* is a logarithmic measure of the observed intensity of a galaxy. If all galaxies had the same luminosity, their observed intensities would fall off as the inverse square of their distances and so their apparent magnitudes would be proportional to the logarithm of the distances of the galaxies. The logarithm of the recessional velocity of the galaxy is plotted along the vertical axis. It can be seen that the points lie beautifully along a straight line at 45° to the axes, which is what would be expected if the recessional velocity is precisely proportional to distance. Correlations never come any better than this in cosmology. All extragalactic objects – galaxies, radio galaxies, quasars, clusters of galaxies – obey the same velocity–distance relation. The traditional way of writing Hubble's law is

$$v = H_0 r$$

where v is the recessional velocity of the galaxy and r is its distance from our Galaxy. H_0 is, appropriately, known as *Hubble's constant*.

To interpret Hubble's law, we need one further piece of information and that concerns the question of whether or not our Universe looks the same in all directions on the largest angular scales. In Chapter 1, we agonized a great deal about the fact that the distribution of galaxies is highly irregular – there are large holes or voids, as well as sheets and filaments made up of galaxies. The question is, if we take large enough regions of the Universe, does the distribution of galaxies eventually end up being smooth, or are there still irregularities on the very largest scales? Figure 1.16 shows the distribution of galaxies in the northern galactic hemisphere and, if regions away from the edge of the picture are studied, it can be seen that, despite the irregularities, one bit of the picture looks very much like another. Thus, on large enough scales, the distribution of galaxies seems to have the same average properties wherever we look. Their distribution is not uniform on the small scale but it has the same degree of irregularity within any large volume we choose. If we think in terms of the sponge model, which we introduced in Section 1.2, there are filaments and sheets as well as large holes, if we look at the sponge on a small scale. When we consider large pieces of the sponge, however,

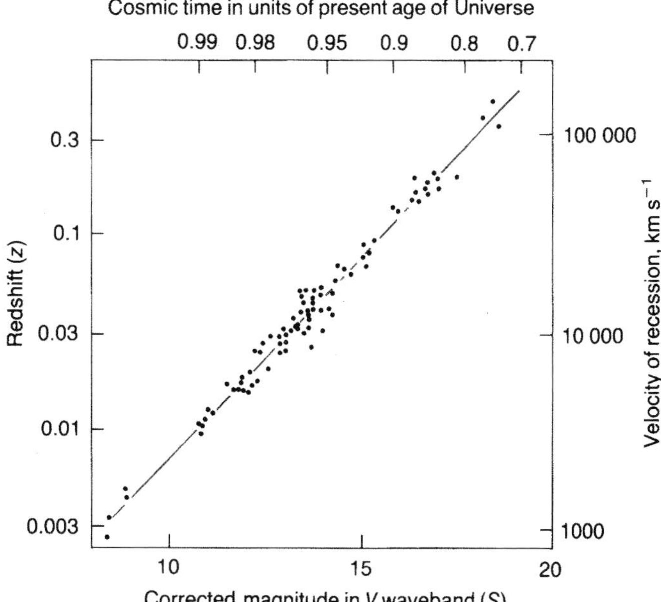

Figure 4.4. A modern version of what is often called the 'Hubble diagram' for the brightest galaxies in clusters of galaxies. In this logarithmic plot, the corrected apparent magnitude V is plotted against redshift, z. The apparent magnitude is defined as $V = \text{constant} - 2.5 \log S$ where S is the flux density, or observed intensity, in the V waveband. The straight line is the relation which would be expected if $S \propto z^{-2}$. The small dispersion about this line shows that the brightest galaxies in clusters must have remarkably standard properties and that the distances of the galaxies are proportional to their recessional velocities or redshifts.

one bit of it looks very much like any other. The Universe of galaxies has exactly the same property.

Nowadays, there is much more evidence about the uniformity of the distribution of matter and radiation on the very largest scales. We can, for example, select only very distant objects by studying large samples of bright radio sources, which we discussed in Chapter 3. Figure 4.5 shows the positions of the brightest 31 000 radio sources, which are listed in the *Greenbank Catalogue of Radio Sources*, in the northern sky. The coordinates have again been distorted so that equal areas on the sky correspond to equal areas on this flat two-dimensional

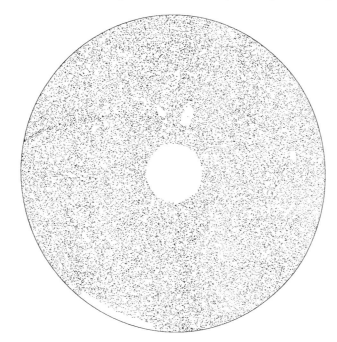

Figure 4.5. The distribution over the sky of the brightest 31 000 radio sources at a wavelength of 6 cm in the Northern Celestial Hemisphere from the *Greenbank Catalogue of Radio Sources*. In this equal area projection, the North Celestial Pole is in the centre of the diagram and the celestial equator around the circumference. The region about the North Celestial Pole was not surveyed. There are 'holes' in the distribution of sources about the bright radio sources Cygnus A and Cassiopeia A and a small excess of sources associated with the Galactic plane. Otherwise, the distribution of sources does not display any deviation from a random distribution.

picture. There is a prominent hole in the centre of the distribution because that area of sky was not surveyed. If we exclude the central hole and the two small holes about the very bright radio sources Cygnus A and Cassiopeia A, the distribution of the points is entirely consistent with the radio sources being located randomly on the sky. Any apparent 'constellations' or clusters of radio sources are simply due to the random superposition of points. We know that the radio sources plotted in Figure 4.5 are radio galaxies and radio quasars and these probe the very largest scales we can study in the Universe at the present day. The uniformity of the distribution of points in Figure 4.5 strongly suggests that the distribution of distant objects is the same in all directions on the sky.

We have, however, even better information about the large scale uniformity of the Universe from the distribution of the cosmic microwave background radiation over the sky. As we will discuss in more detail in Chapter 5, this radiation is the cooled remnant observable today of the hot early phases of the Universe. We have already discussed the very beautiful millimetre map of the sky obtained by the COBE satellite (Figure 1.29) in Chapter 1. That picture was dominated by the slight increase and decrease in the intensity of the radiation in opposite directions on the sky associated with Earth's motion through a frame of reference in which the intensity of the microwave background radiation would be the same in all directions. The COBE scientists have taken extraordinary pains to remove the effects of the Earth's motion upon the intensity distribution over the sky and have created the remarkable map of the millimetre sky shown in Figure 4.6. The plane of our own Galaxy now appears very prominently as an intense band of emission across the centre of the diagram. Away from the

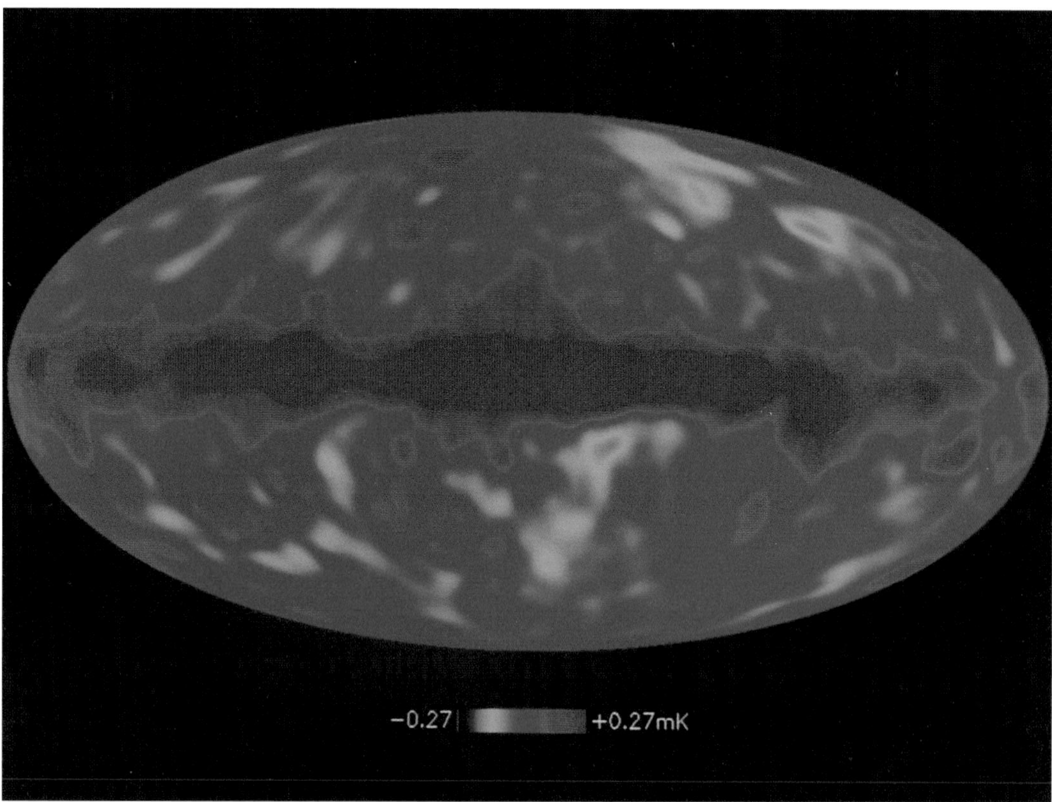

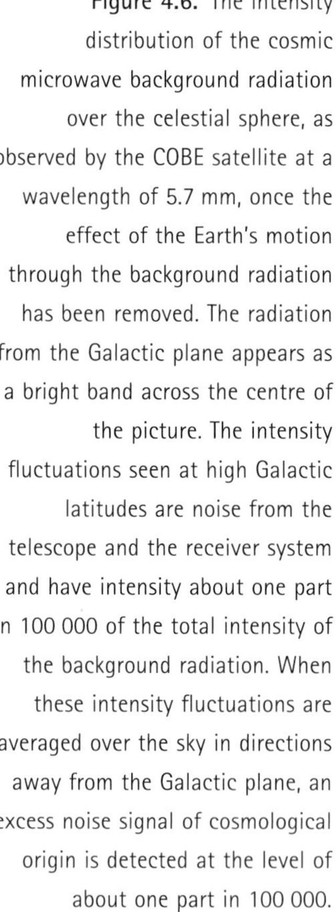
Figure 4.6. The intensity distribution of the cosmic microwave background radiation over the celestial sphere, as observed by the COBE satellite at a wavelength of 5.7 mm, once the effect of the Earth's motion through the background radiation has been removed. The radiation from the Galactic plane appears as a bright band across the centre of the picture. The intensity fluctuations seen at high Galactic latitudes are noise from the telescope and the receiver system and have intensity about one part in 100 000 of the total intensity of the background radiation. When these intensity fluctuations are averaged over the sky in directions away from the Galactic plane, an excess noise signal of cosmological origin is detected at the level of about one part in 100 000.

Galactic plane, small intensity fluctuations can be observed but these are at a level of only one part in 100 000 of the total intensity of the cosmic microwave background radiation. What this means is that the intensity of the cosmic microwave background radiation is the same in all directions to an accuracy of better than one part in 100 000. This is quite remarkable precision for any cosmological experiment.

One might ask, 'Well, what has the cosmic microwave background radiation to do with the Universe of galaxies?' – we will address this question later in this chapter. What we will show is that this radiation provides direct evidence about the smoothness of the distribution of matter in the Universe long before the galaxies formed. The fact that the radiation is so incredibly uniform in all directions tells us that the matter was also very smoothly distributed in all directions before the galaxies formed. Therefore, when we construct cosmological models for the dynamics of the Universe as a whole, it is a very good approximation to begin with uniform models which are the same in all directions. Technically, we say that the Universe is *isotropic* and homogeneous on the large scale.

When we combine Hubble's law with the remarkable isotropy of the Universe, we find that it has a deeper meaning. Figure 4.7 illustrates a distribution of galaxies expanding uniformly. What we mean by a *uniform expansion* is that, if the galaxies were placed on a uniform grid, as shown in Figure 4.7, the distances between neighbouring galaxies would all increase by the same amount in the same time interval. The trick is now to concentrate attention upon any one of the galaxies and ask how the other galaxies are observed to move relative to the chosen galaxy. It is simplest to concentrate upon a line of galaxies, such as the galaxies A, B and C in Figure 4.7. It can be seen that, in the uniform expansion, the further away a galaxy is from the chosen galaxy, the further it has to move in a given interval of time

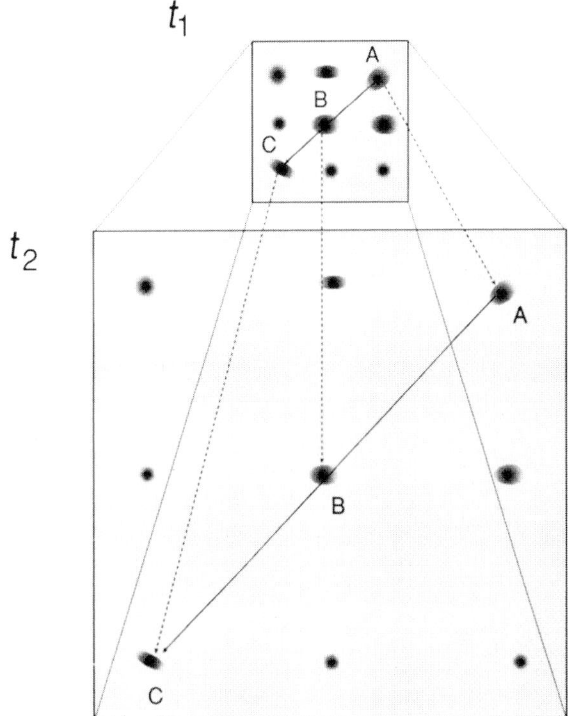

Figure 4.7. Illustrating the origin of Hubble's law for a uniform distribution of galaxies partaking in a uniform expansion. The system of galaxies expands uniformly between the epochs t_1 and t_2. If we consider the motions of galaxies relative to galaxy A, it can be seen that galaxy C travels twice as far as galaxy B and so has twice the observed recessional velocity of galaxy B, relative to A. Since C is twice the distance of B, it can be seen that Hubble's law is a general property of isotropically expanding Universes.

in order to preserve the uniformity of the expansion. By inspection of Figure 4.7, it can be seen that, in a uniform expansion, the observed recessional velocity of a galaxy from a chosen galaxy is precisely proportional to its distance, that is, Hubble's law. It can be demonstrated rigorously that, in a uniformly expanding Universe, the further away a galaxy is, the faster it has to be moving away from us in order to guarantee that the distribution of galaxies remains isotropic and uniform. The other remarkable aspect of this demonstration is that all observers partaking in the uniform expansion observe the distribution of galaxies to be expanding uniformly about their own positions and they all measure the same value for Hubble's constant. This is the real significance of Hubble's law – it tells us that the Universe as a whole is expanding uniformly. All observers note, correctly, that they are at the 'centre of the Universe' and, when the expansion is extrapolated back to the origin, they were all present at a single point at the very beginning of the expansion.

One of the immediate consequences of this result is that the Universe must have started off very compact and very hot because, as in the explosion of a supernova, matter cools as it expands – in the same way, matter and radiation cool down in the expanding Universe. The early phases of the Universe must therefore have been very hot indeed and that is why this model is called the *Big Bang* model of the Universe, which is the subject of Chapter 5. In this chapter, we restrict attention to the problem of forming galaxies in the expanding Universe.

Before we return to that problem, let us consider how we expect the Universe to expand under the influence of its own gravity. Besides its importance for understanding the dynamics of the expanding Universe, we will find that this analysis provides insights into the origin of the problems of forming galaxies.

4.3 The dynamics of the expanding Universe

The considerations of the last section suggest that the obvious starting point for the construction of models of the Universe as a whole is to consider uniform, isotropic and uniformly expanding models. The question we want to answer is 'How does the rate of expansion of the Universe change with time?' In other words, what are the dynamics of the expanding Universe? The first step is to smooth out all the structure we observe in the real Universe and to replace it by a completely smooth distribution of matter which expands uniformly. This may seem a rather drastic step but it is justified by the spectacular isotropy of the cosmic microwave background radiation, which, as we argued in the last section, shows that the Universe is remarkably isotropic on the large scale. The fact that these models are uniform and isotropic simplifies the calculations enormously.

We can think of the dynamics of the models in terms of a competition between the uniform expansion, which is dispersing all the matter in the Universe, and the force of gravity which is trying to prevent that happening, that is, to pull all the matter back together again. It turns out that we can model the dynamics of the Universe precisely if we ask a simpler question, 'What is the deceleration due to gravity of a galaxy located at the surface of a uniformly expanding sphere, which has density equal to the average density of the Universe?' The calculation is illustrated in Figure 4.8. This simplification works because every observer in the Universe could carry out exactly the same calculation and find exactly the same answer – each of them observes exactly the same isotropically expanding Universe, as we demonstrated in Figure 4.7. The same dynamics come out of this calculation as are found in the full

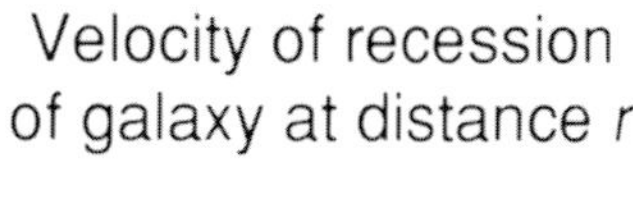

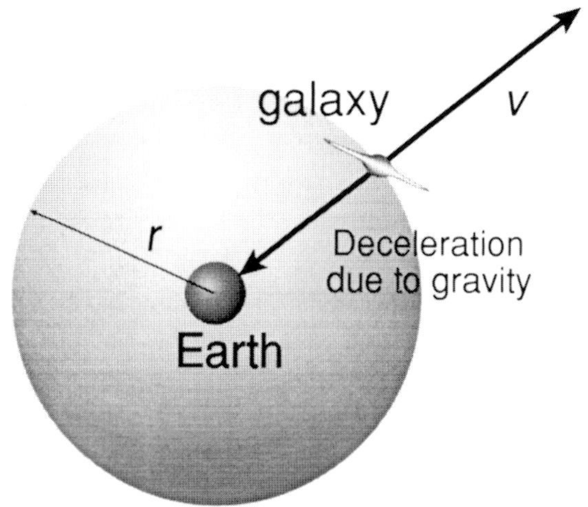

Figure 4.8. Illustrating a simple Newtonian model for the dynamics of isotropic world models. The galaxy at distance r is decelerated by the gravitational attraction of the matter within distance r of our own Galaxy. Because of the assumption of isotropy, an observer on any galaxy participating in the uniform expansion would carry out exactly the same calculation.

theory of the expanding Universe according to the general theory of relativity. The reason for this is that, in isotropic models of the Universe, local physics is also global physics – in other words, the properties of each small region of the Universe must also contain information about its global structure.

The general solutions for the dynamics of the standard world models of general relativity were discovered by the Russian meteorologist and theoretical physicist Alexander Friedman in the years 1922 to 1924 and they are illustrated in Figure 4.9. In these models, the dynamics depend entirely upon the average density of matter. If the Universe is of high density, the gravitational deceleration acting upon each piece of matter can be sufficiently great to halt the expansion of the Universe which then collapses back to a hot dense phase – this event is sometimes referred to as the *Big Crunch*. If the Universe is of low density, there may be insufficient mass present in the Universe to prevent the matter expanding to infinity. In this case, the Universe ends up being of infinite size and it is still expanding with a finite velocity as

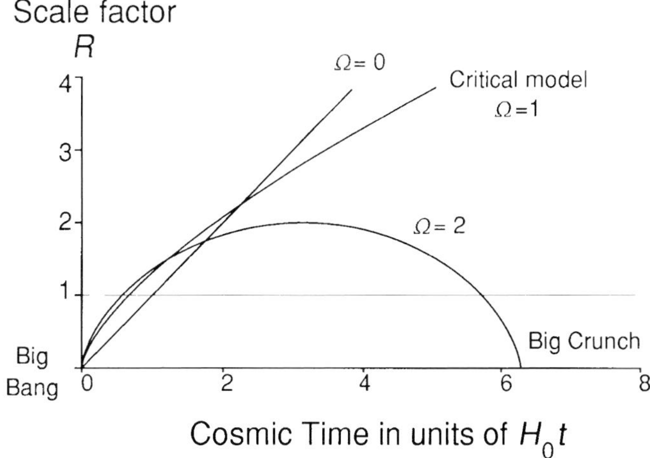

Figure 4.9. A comparison of the dynamics of different Friedman models of the expanding Universe, characterised by the density parameter $\Omega = \rho/\rho_{crit}$. For convenience, the time coordinate, which is called *cosmic time*, is measured in units of $H_0 t$, where H_0 is Hubble's constant, and the scale factor takes the value $R = 1$ at the present time. In this presentation, the different world models all have the same value of Hubble's constant at the present epoch, that is, the tangents to the curves all lie at an angle of 45° to the axes at $R = 1$. If $\Omega > 1$, the Universe collapses to a Big Crunch, $R = 0$, at some point in the future. If $\Omega < 1$, the Universe expands to infinity and has a finite velocity of expansion when it gets there. The critical model has $\Omega = 1$ and just expands to infinity and no more. The advantage of this presentation is that time-scales for these world models can be easily derived using the value of Hubble's constant preferred by the reader.

the Universe becomes infinitely old. Between these two types of behaviour, there is what is known as the *critical* or *Einstein–de Sitter model.* This model just expands to infinity and no more. In other words, the velocity of expansion of the Universe goes to zero when the Universe is of infinite age. This model is of special importance for cosmology as we will discover in Chapter 5.

The density of the Einstein–de Sitter model is known as the *critical density*, ρ_{crit}, and this is a very important reference quantity in cosmology which we will encounter over and over again. Of course, the Universe may actually have some other average density ρ, which is quite different from ρ_{crit}. It is then convenient to introduce what is known as the *density parameter*, Ω, which is just the ratio of the actual mean density of the Universe ρ to the critical density, ρ_{crit}, that is, $\Omega = \rho/\rho_{crit}$. Thus, models with $\Omega > 1$ eventually collapse to a hot dense end while those with $\Omega \leq 1$ expand forever. The dynamics of three of these models are shown in Figure 4.9 – the critical model with $\Omega = 1$, a model with $\Omega = 2$ which collapses to a Big Crunch and a completely empty model which has $\Omega = 0$ and which expands at a constant rate to infinity. In Figure 4.9, the expansion of the Universe is described in terms of the *scale factor R* which describes how the relative separation of any two points, which partake in the uniform expansion of the Universe, changes with time. For convenience, the scale factor has been taken to have the value 1 at the present epoch. Thus, in the past, the scale factor had values less than 1 and will have values greater than 1 in the immediate future.

There is an analogy between the dynamics of the Friedman world models and the concept of *escape velocity*, which is illustrated in Figure 4.10. A spaceship acquires a certain velocity at the surface of the Earth and it only escapes from the pull of the Earth's gravity, if this velocity is large enough. If the velocity of the spaceship exceeds the escape velocity, it escapes from the pull of Earth's gravity and can reach an infinite distance from the Earth. If it has less than the escape velocity, it does not escape from the pull of the Earth's gravity and falls back to Earth. Between these extremes, there is an initial velocity which corresponds to the spaceship just escaping to infinity and no more – this velocity is known as the escape

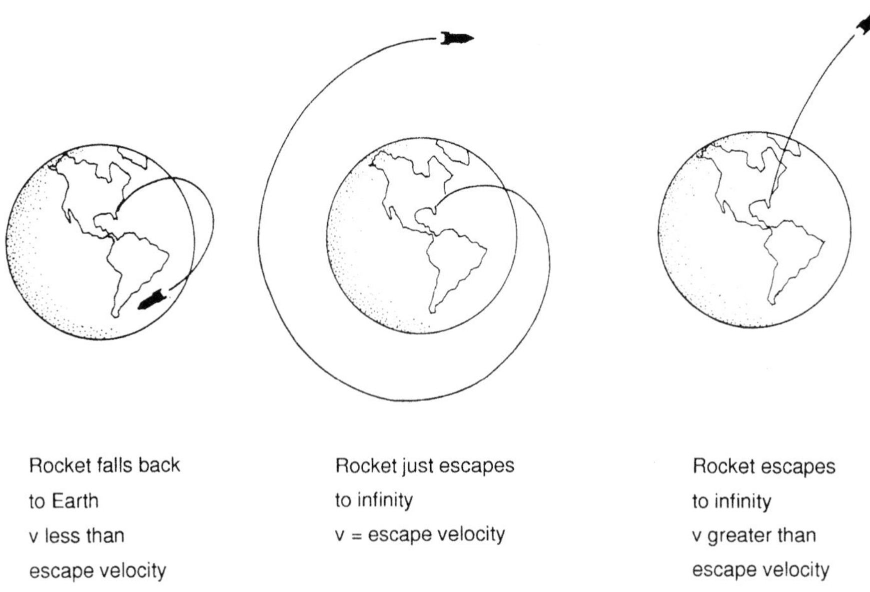

Figure 4.10. Illustrating the concept of escape velocity for a space rocket launched from the surface of the Earth.

velocity. These three types of behaviour are analogous to the dynamics of Universes which have densities less than, greater than and equal to the critical density, Ω_{crit}, respectively.

In the Friedman world models, the deceleration of the Universe as a whole is entirely determined by the average mass density of the Universe now. There is one other interesting and important feature of these models of the Universe and that is that there is a very close relation between the geometry of the Universe and its average density. This is an aspect of the models which cannot be included in the simple Newtonian model, but which is automatically built into Einstein's general theory of relativity. As discussed in Section 3.6, Einstein's theory describes how matter bends the geometry of space–time and so, in general, we expect that the geometry of space will not be the flat Euclidean geometry with which we are familiar in everyday life. There is one important result which we will need in Chapter 5 concerning the geometry of space and so let us discuss this point in a little more detail. If we draw a triangle on a flat sheet of paper, the angles at the corners of the triangle add up to 180°. This is a property of flat space. Now suppose we draw the triangle on the surface of a sphere. Suppose, for the sake of definiteness, that we draw a triangle from the 'north pole' of the sphere down to the equator and that the angle between the lines at the north pole is 90° (Figure 4.11). We complete the triangle by drawing a line along the equator to meet the lines drawn from the north pole. Notice that all three angles of the triangle are 90° and so the sum of the angles of this triangle is 270°. This is an example of curved space – the sum of the angles of the triangle does not add up to 180°. This is what the presence of matter does to the geometry of space. The geometry of space is slightly bent and, in principle, this can be measured.

In the standard models of the Universe, there is a beautifully simple relation between the geometry of space and the average density of matter in the Universe. If the density of the

Figure 4.11. Illustrating the difference in the sum of the angles of a triangle if the triangle is drawn in (*a*) flat space and (*b*) on the surface of a sphere.

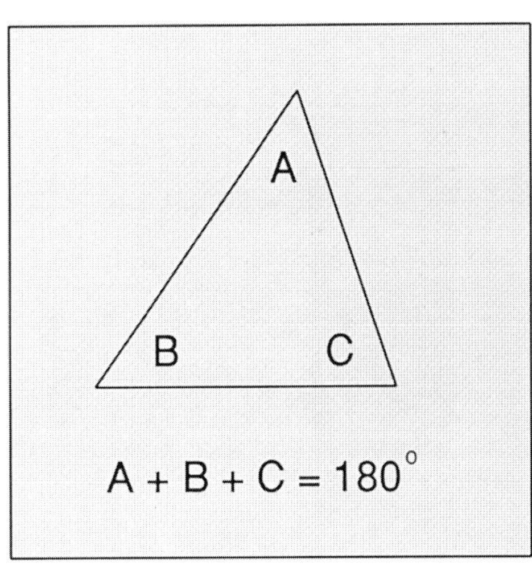

(*a*)

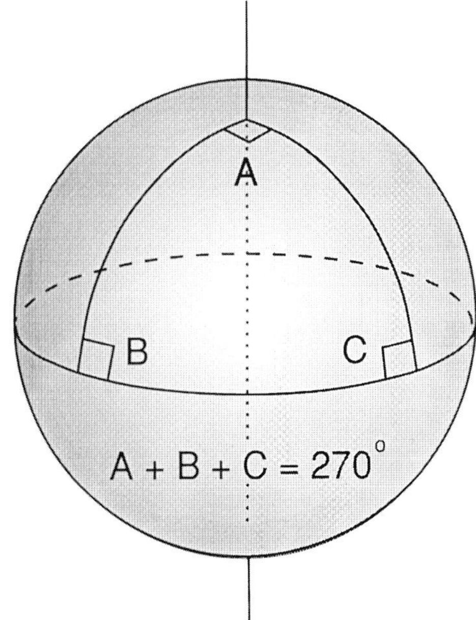

(*b*)

Universe exceeds the critical density, the geometry is spherical, if it is less than the critical density, the geometry is hyperbolic and only in the case of the critical model, $\Omega = 1$ is the geometry Euclidean. We need not worry about the curvature of space locally, because it only becomes important on the scale of the Universe. Therefore, the angles of a triangle add up to 180° very precisely in our locality. If, however, our sheet of graph paper extended out to what we called the 'size of the Universe at the present day' in Chapter 1, then, the angles of a triangle with sides on that scale might not add up to 180° – indeed, in general, they will add up to some different value. We will need this concept of the relation between geometry and the mass density of the Universe later at a crucial point in the discussion of the early Universe.

Another important part of this story concerns the relation between the dynamics of the Universe and its mass density ρ. In the standard models, the Universe decelerates under the gravitational influence of its matter content and the deceleration is exactly proportional to the average density of matter in the Universe today. We can therefore test the validity of the models by comparing the rate at which the Universe is decelerating with its average mass density and find out whether or not they match up – that is, can we attribute the observed deceleration of the Universe entirely to the mass which we know to be present? If we were able to show that this is indeed the case, we would have obtained a remarkable validation of Einstein's theory of general relativity on the largest scales we can study in the Universe.

The way in which this comparison is made in the professional literature is in terms of the parameters which describe the density and deceleration of the Universe. We have already introduced the *density parameter* Ω. We can also define the deceleration of the Universe in terms of a quantity known as the *deceleration parameter*, which is labelled q_0. The deceleration parameter is defined to be minus the rate at which Hubble's constant changes with time divided by the square of Hubble's constant, both quantities being measured at the present time. The important point is that, according to the standard world models, $\Omega = 2q_0$. Therefore, the critical density $\Omega = 1$ corresponds to a value of the deceleration parameter $q_0 = \frac{1}{2}$. What is important is that the present deceleration of the Universe and its average mass density are independently measurable quantities and so provide a test of the general relativistic models of the Universe. Let us see how well this comparison can be made.

4.4 The measurement of cosmological parameters

The pioneers of cosmology of the 1930s realised that which Friedman model of the Universe we live in could be determined by measuring Hubble's constant H_0, the deceleration parameter q_0 and the density parameter Ω. Indeed, by comparing q_0 and Ω, the validity of the simplest isotropic models of general relativity could be tested on the largest scales accessible to us in the Universe. This has turned out to be a much more difficult task than these pioneers anticipated. Let us discuss briefly the current status of these determinations.

4.4.1 Hubble's constant and the age of the Universe Hubble's original estimate of the value of Hubble's constant was about 500 km s^{-1} Mpc^{-1} and this raised problems for the simplest Friedman models. The age of the Universe is related to the inverse of Hubble's constant, which, for convenience, we will write as $T_0 = 1/H_0$. Let us illustrate this by considering the simple case of the empty model, $\Omega = 0$. In this model, there is no deceleration of test particles because there is no matter present in the Universe. Therefore, the Universe must have expanded at the same constant rate since the Big Bang and so the scale factor, R, changes with time as $R = H_0 t$ (see Figure 4.9). If we write the present age of the Universe as t_0 and set the scale factor of the Universe at the present day equal to one, we see that $t_0 = T_0$. This can be confirmed by consulting Figure 4.9, in which it can be seen that the present age of the Universe for the $\Omega = 0$ model is $1/H_0$. It can also be seen that the ages of the simple Friedman models with $\Omega > 0$ all have ages less than T_0. For example, the critical model, which has $\Omega = 1$, has age $t_0 = \frac{2}{3}T_0$. If Hubble's constant were 500 km s^{-1} Mpc^{-1}, T_0 would be only 2×10^9 years and so the age of the Universe would have to be less than this value. This age was in clear conflict with the age of the Earth, the present best estimate being 4.6 billion years. To remedy this problem, more complex Friedman solutions of Einstein's equations were adopted which included a term called the *cosmological constant*, λ – this had the effect of adding a repulsive force to counteract the attractive force of gravity. The result was that the cosmological time-scale could be increased to arbitrarily large values, but at the expense of adding an additional complication to the simple picture of a Universe evolving only under the influence of gravity. Examples of these types of model are illustrated in Figure 4.12 and they were advocated by Eddington and Lemaître as a solution to the problem of the cosmological time-scale.

In the 1950s, however, the value of Hubble's constant was continually revised downwards as better understanding of the cosmological distance measures was gained. In 1952, Walter Baade revised the value of Hubble's constant downwards to 250 km s^{-1} Mpc^{-1} and then in 1956 Sandage revised it down yet again to 75 km s^{-1} Mpc^{-1}. These revisions eliminated the need for the cosmological constant. Since that time, however, the precise value of Hubble's constant has been a matter of controversy, one school favouring values close to 100 km s^{-1} Mpc^{-1} and the other values close to 50 km s^{-1} Mpc^{-1}. The cause of these discrepancies is the difficulty of measuring extragalactic distances which are independent of the redshifts of the galaxies. The traditional methods involve identifying the same types of object in nearby and distant galaxies and then measuring relative distances by measuring their apparent magnitudes. This procedure is, however, very sensitive to small, subtle corrections and to systematic biases in the data unless great care is taken in the reduction and analysis of the data.

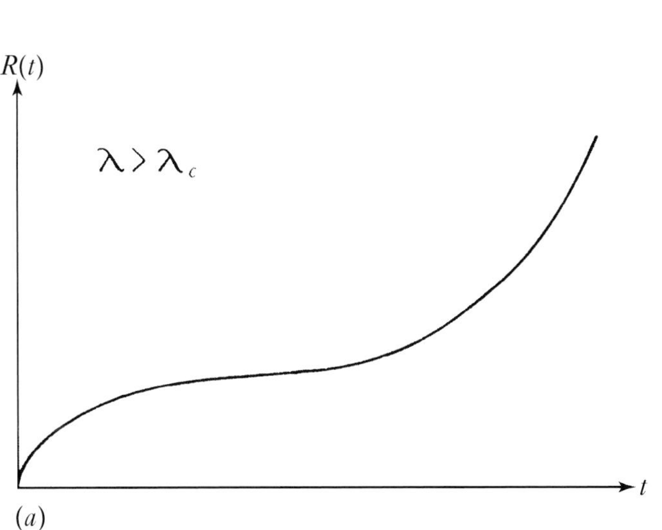

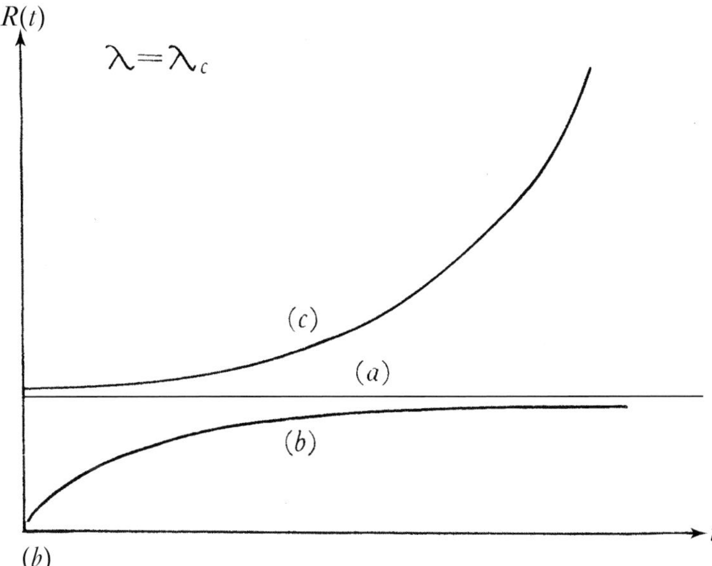

Figure 4.12. Examples of the dynamics of world models which include the cosmological constant. The scale factor on the vertical axis represents the separation between any two typical points, which partake in the universal uniform expansion. The effect of the repulsive force associated with the cosmological constant is that the time-scales of the standard models can be stretched to values much greater than $1/H_0$. (*a*) A Lemaître model with positive cosmological constant. (*b*) There is one special case of these models, known as the Eddington–Lemaître model, in which the Universe was stationary with the repulsive effect of the cosmological constant being exactly balanced by the attractive force of gravity. This static model is represented by the line marked (*a*). The line (*b*) indicates how the stable state is reached by expansion from the infinitely distant past: the line (*c*) shows how the model evolves, if perturbed away from the static state. Models of this type can have infinite age.

The most recent analyses have used observations of Cepheid variable stars in galaxies in the Virgo cluster of galaxies, the most beautiful data set being obtained for the galaxy M100 from observations made by the Hubble Space Telescope. The Cepheid variables belong to a special luminous class of regular variable star with a distinctive light curve which can be readily detected by the Hubble Space Telescope in galaxies in the Virgo cluster (Figure 4.13). The property of the Cepheid variables which makes them so useful as distance indicators is that there is a very well defined relation between their periods and their luminosities. The longer their periods, the more luminous they are. This relation is so well defined that, by measuring the period of a Cepheid variable, its intrinsic luminosity can be estimated and so, by measuring its apparent magnitude, its distance can be found. The Hubble Space Telescope observations are very beautiful indeed and, in the galaxy M100, 12 Cepheids have been found, which follow precisely the expected period–luminosity relation. These observations have provided an estimate for Hubble's constant of 80 ± 17 km s^{-1} Mpc^{-1}. This is the first result of a programme which will involve the measurement of Cepheid variables in a number of galaxies in the Virgo cluster. These observations are expected to reduce considerably the uncertainty in the value of Hubble's constant.

Taken at face value, it looks as if the cosmological time-scale problem of the 1930s and 1940s has returned to haunt us. As we discussed in Section 2.3, the ages of the oldest stars are currently estimated to be about 14 to 16 billion years, whereas, for $H_0 = 80 \pm 17$ km s^{-1} Mpc^{-1}, T_0 is $(12 \pm 3) \times 10^9$ years. It is important to note that the errors quoted refer to what are known as the '1-σ uncertainties' and describe the statistical uncertainty in the estimate of H_0. Formally, these errors mean that there is a 68% probability that the true value of Hubble's constant lies between 63 and 97 km s^{-1} Mpc^{-1} but there is a 32% chance that the true value lies outside these limits. Thus, interpreted literally, the value of T_0 is consistent with the ages of the oldest stars. There may, however, be a discrepancy with the age of the critical world model, which would be $t_0 = \frac{2}{3}T_0 = (8\pm2) \times 10^9$ years. It is too early yet to say whether or not it will be necessary to reintroduce the cosmological constant in order to reconcile the age of the oldest stars with the simplest critical Friedman model. We need many more measurements of the quality of the new Hubble Space Telescope data to obtain a secure estimate.

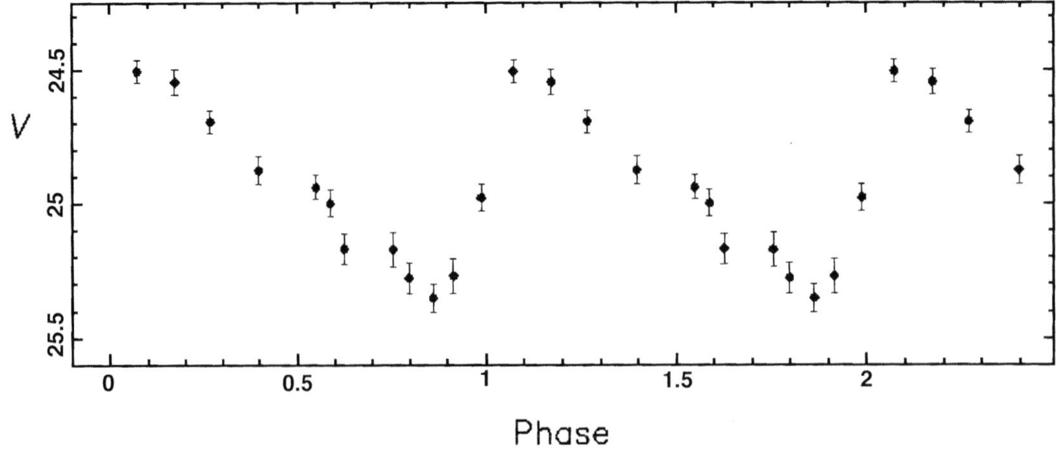

Figure 4.13. An example of the light curve of one of the Cepheid variable stars discovered by the Hubble Space Telescope in the galaxy M100, which is a member of the Virgo cluster of galaxies. The characteristic signature of the Cepheid variables is that they are regular variable stars in which the luminosity increases rapidly followed by a slow decline.

Fortunately, more physical methods of estimating Hubble's constant have been developed and these enable some of the problems of identifying the same types of object in different types of galaxy to be eliminated. I have a feeling, however, that this story is going to run and run.

4.4.2 *The deceleration parameter and the problem of cosmological evolution* It was noted by the early pioneers of observational cosmology that the observed intensity of a distant galaxy at a given redshift depends upon the choice of world model and, in particular, upon the deceleration of the Universe. Inspection of Figure 4.9 illustrates how this comes about. An important result, which we will return to in Section 4.8, is that the scale factor R is directly related to the redshift, z, of a galaxy by the simple relation

$$R = \frac{1}{1+z}.$$

Therefore, a given redshift corresponds to a line at constant R across Figure 4.9 and we can work out from that diagram the time it takes light to travel from the source to the observer at $R = 1$ by comparing the time of emission t corresponding to R with the present age of the model t_0. The curvature of the scale factor–cosmic time relations means that the biggest time difference between the time the light was emitted at redshift z and the time the observer at $z = 0$ receives the light occurs in the empty model which has $q_0 = \Omega = 0$. The greater the deceleration, corresponding to a greater value of Ω, the smaller the time difference. Therefore, at a given redshift, we would expect a standard galaxy to be brighter in world models with $q_0 > 0$ as compared with the $q_0 = 0$ model – the greater q_0, the brighter the galaxy. The effect only becomes important at reasonably large redshifts, $z \approx 0.5$, and the problem has been to find suitable standard galaxies at large redshifts with which to perform the test. Notice that it is traditional to work in terms of the deceleration parameter q_0 in studying the differences in observed intensity between the world models because these are primarily associated with the different kinematics of the world models. We will use q_0 in this section, although we note that, for the simplest Friedman models, $2q_0 = \Omega$.

The brightest galaxies in clusters have a well defined redshift–apparent magnitude relation (Figure 4.4) but it only extends to redshifts of about 0.5 at which the differences between the world models are small. Simon Lilly and I found another way of approaching the problem using the galaxies associated with luminuous radio sources, similar to that associated with the radio source Cygnus A (Figure 3.2). We found that the redshift–apparent magnitude relation in the near infrared waveband at 2.2 μm could be extended out to redshifts of 1.5, at which there is a large difference between the predicted apparent magnitudes of galaxies of the same intrinsic luminosity for different values of q_0. Remarkably, we found that these radio galaxies have a very narrow dispersion in apparent magnitude about the mean relation and so the redshift–magnitude relation is well defined out to redshifts of 1.5. A comparison of our observations with the expectations of the standard world models is shown in Figure 4.14. It can be seen that our observed relation differed from the expectations of world models with values of q_0 of 0 and $\frac{1}{2}$. The distant galaxies are much brighter than they should be. We interpreted this as being due to the fact that the galaxies were systematically brighter in the past, as would be expected since the stars were evolving off the main sequence onto the giant branch at a much greater rate in the past as compared with the present day. In fact, the simplest models of the evolution of these galaxies suggested that the galaxies should be about a

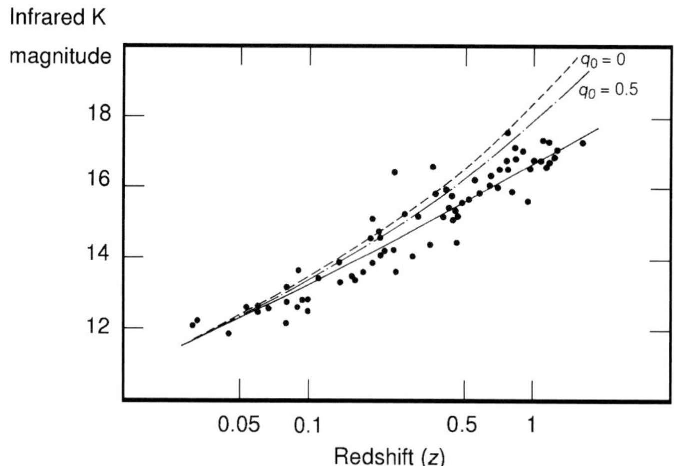

Figure 4.14. The redshift–apparent magnitude relation measured in the infrared waveband at 2.2 μm for radio galaxies in the 3CR catalogue. The observed relation extends out to redshifts of about 1.5 and the distribution of the radio galaxies about the mean line remains narrow from small to large redshifts. The expectations of uniform world models with $q_0 = 0$ and $\frac{1}{2}$ are also shown. It can be seen that the galaxies at redshift 1 are about 1 magnitude brighter than would be expected according to the standard world models. In our analysis of these data, we attributed this difference to the evolution of the stellar populations of the galaxies which would make them about a magnitude more luminous at a redshift of 1 compared with their luminosities at the present epoch.

magnitude brighter at a redshift of 1 as compared with their luminosities at the present day. This evolutionary change can account for the difference between the observations and the expectations of the standard world models.

This example illustrates one of the fundamental problems of trying to determine the value of q_0 from studies of distant objects. In order to observe a significant difference between the world models, it is necessary to study objects at large redshifts but then they are observed at times when the Universe was much younger than it is now and we have to take account of the evolutionary changes which may have taken place over that period. Thus, until we can understand the astrophysics of the objects used in these cosmological tests, we do not know how to make allowance for the evolutionary changes which may have taken place. In the example described above, the best we could obtain was an estimate of $q_0 = 0.5 \pm 0.5$. There might well be other systematic trends which we do not yet understand. Equally serious, in my view, is the problem that we do not have a good understanding of why these radio galaxies all have roughly the same intrinsic luminosities. We really ought to have a secure understanding of the astrophysics of these objects before they are used in cosmological tests. Thus, the value of q_0 is not at all well known from the classical cosmological tests, which is a disappointment after so much effort by many astronomers.

4.4.3 *The density parameter and the dark matter problem* In many ways, we have better knowledge of the density parameter than of the deceleration parameter because we can measure the mass density of the Universe in galaxies and clusters and so obtain at least lower limits to the mass density of the Universe. This type of calculation was first carried out by Hubble in 1926 as soon as the extragalactic nature of the spiral nebulae was established. The mass contained within the visible parts of galaxies can be estimated and then, knowing their space density, the total mass density attributable to galaxies can be estimated. Performing this calculation, it is generally agreed that the mass contained within the visible parts of galaxies amounts to only about 1% of the critical density. This is where we encounter a fundamental problem – we are now certain that most of the mass in the Universe is not contained within the visible parts of galaxies. In addition to the visible matter, there must also be *dark matter*. This is such an important topic that it deserves a section to itself.

4.5 The dark matter problem

One of the most embarrassing admissions, which astronomers and cosmologists have to make, is that they are not really certain what form most of the mass in the Universe takes. It is perhaps rather remiss to have left this admission until Chapter 4. The problem arises as follows. The masses of galaxies and clusters of galaxies are determined using essentially the same techniques discussed in connection with the measurement of the masses of the supermassive black holes in the nuclei of NGC 4151 and M87 (Section 3.8). The centrifugal forces acting on the stars or galaxies are balanced by the gravitational forces of the systems as a whole and, from this, the masses of the systems can be estimated. When this calculation is carried out for giant spiral galaxies, such as the edge-on spiral galaxy NGC 5084 (Figure 4.15), it is found that the density of matter falls off much more slowly than the light of the galaxy with increasing distance from its centre. There must therefore be a large amount of *dark matter* in the outer regions of the galaxies which emits very little light. Typically, there must be about ten times more mass present than would be inferred from the light of the galaxy.

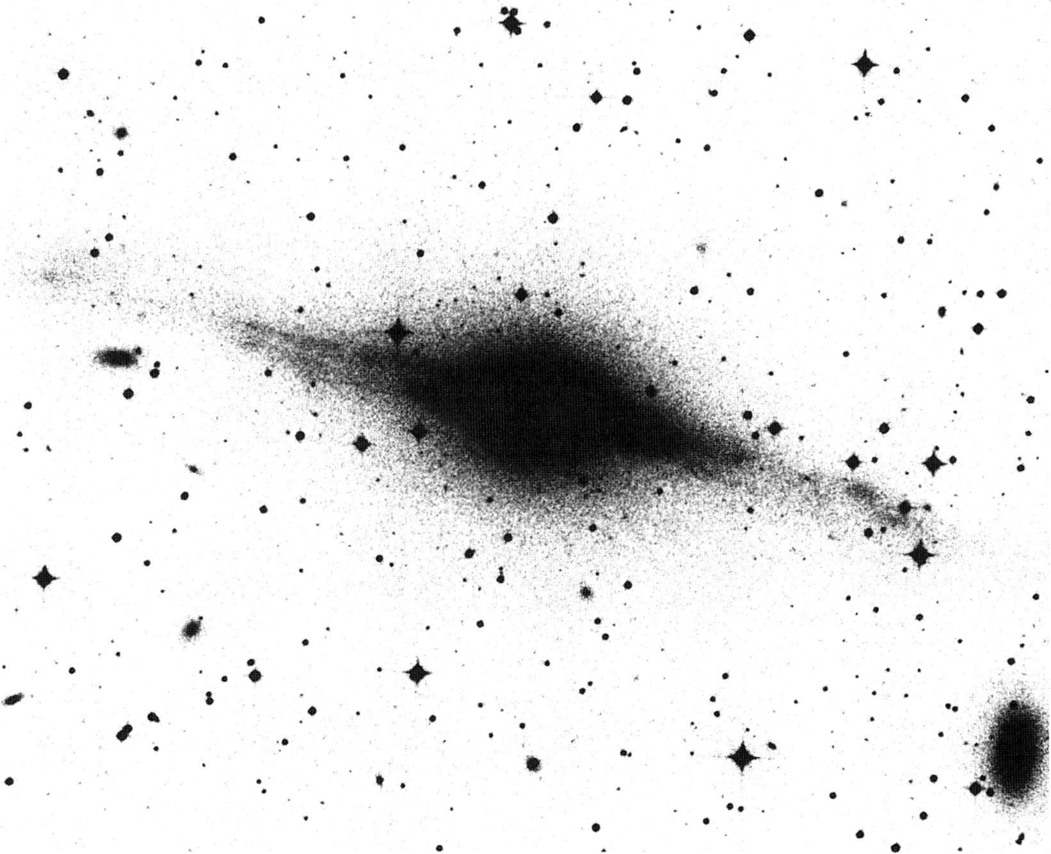

Figure 4.15. The edge-on giant spiral galaxy NGC 5084, which is known to contain about ten times as much dark matter as visible matter.

Further evidence for the presence of dark matter is found in giant clusters of galaxies such as the Pavo cluster (Figure 1.15). This is by no means a new realisation, the first analysis of the problem having been carried out by Fritz Zwicky in 1937. By measuring the velocities of the galaxies in a cluster, its total mass can be estimated and is found to exceed by a factor of about 10 to 20 what would be expected on the basis of the light of the galaxies in the cluster. There may well be much more dark matter in the Universe on even larger scales. There is evidence for bulk streaming motions of galaxies on a large scale and also for deviations from the smooth Hubble expansion caused by the gravitational influence of superclusters of galaxies. The amount of mass needed to cause these motions exceeds by a large factor that which could be contained within the visible parts of galaxies. These different pieces of evidence constitute what is known as the *dark matter problem*. The key questions are:

- How much dark matter is there in the Universe?
- What is the nature of the dark matter?

The first question is the easier of the two. As discussed in Section 4.3, the total mass density of the Universe is a crucial quantity for all cosmology. For our present purposes, the question is how much dark matter there is relative to the mass density of the visible matter and how these densities compare with the critical density, ρ_{crit}, that is, what is the value of the density parameter Ω for ordinary and dark matter? As noted above, the amount of matter present in the visible parts of galaxies amounts to only about 1% of the critical density. We can therefore be confident that those forms of matter which emit light cannot close the Universe. Most estimates of the amount of dark matter relative to visible matter in galaxies and clusters of galaxies suggest that there is about 10 to 20 times more mass in the dark as compared with visible matter. If this result were to apply universally, we could account for about 10 to 20% of the critical density. What is not clear is whether or not, on scales greater than those of clusters of galaxies, there is even more dark matter so that the density parameter would approach unity. Analyses of the velocity field in the local Universe have suggested that this may indeed be the case but it cannot yet be considered firmly established. It is entirely on the cards that our Universe has the critical density but that would lead to the unhappy situation that most of the dark matter has to be located in the places where it is most difficult to measure it, namely, in the spaces between clusters of galaxies.

We can now compare the values of q_0 and Ω. It can be seen that there are still considerable uncertainties in both cosmological parameters. The strongest constraint is that the density parameter Ω must certainly be greater than about 0.01 and I find the evidence convincing that it must be greater than 0.1. It is a key cosmological result that Ω is within a factor of ten of the critical value $\Omega = 1$. Whether or not it is precisely 1 is, in my view, still an open question. The value of q_0 is much more uncertain and within the uncertainties, could even be negative. This would be expected if it were necessary to introduce the cosmological constant into Einstein's equations in order to reconcile the age of the oldest stars with the value of $T_0 = H_0^{-1}$. An optimist might claim that we know that $\Omega = 2q_0$ within about a factor of ten. What is certain is that we urgently need better estimates of both parameters to address this central issue for cosmology and for physics in general.

There is no question but that there is dark matter in the Universe and that it plays a key role in stabilising galaxies and clusters of galaxies but what is it made of? Here is a list of some possible forms for the dark matter:

- Interstellar planets
- Brown dwarfs
- Very low mass stars
- Isolated neutron stars

- Little black holes
- Big black holes
- Supermassive black holes
- Massive neutrinos
- Unknown weakly interacting particles
- Standard bricks
- Abandoned spaceships
- Copies of the *Astrophysical Journal*

The list begins with astrophysically very respectable things like interstellar planets, very low mass stars, brown dwarfs and black holes of different masses, then goes on to some of the exotic types of particle discussed by the particle physicists and ends with really exotic things like bricks, spaceships, copies of the *Astrophysical Journal* and so on. Now, some of the items towards the end of this list may look rather silly but they are there to make an important point. Dark matter could be present in the Universe in any of these forms, and could make significant contributions to its total mass density, but we would not know that they are there. We only know of the presence of different forms of matter in the Universe if they emit detectable radiation, or if they absorb the radiation of background objects. To take one of the more humorous examples in the above list, in order to attain the critical density, there would need to be only one kilogram brick per cube of side roughly 500 million kilometres and, if these were uniformly distributed throughout the Universe, they would not obscure the most distant objects we can observe in the Universe. The bricks could be so cold that they would not emit a significant amount of radiation. This is not to be taken as a serious suggestion, any more than that the missing mass is made up of abandoned spaceships or copies of the *Astrophysical Journal* – the important point of principle is that certain types of matter can be very difficult to detect in the Universe, even if they are very abundant.

Some of these forms of dark matter must be important at some level in the Universe – for example, brown dwarfs, rocks, isolated neutron stars and black holes are certainly present in the Universe. What has excited a great deal of interest among cosmologists and the particle physics community is the possibility that the dark matter may consist of the exotic particles predicted by some of the more promising theories of elementary particles. These particles would interact only very weakly with ordinary matter and are only created at energies far in excess of those attainable in the most powerful terrestrial accelerators. Thus, the particle physicists regard the early Universe as a laboratory for the study of very high energy processes. These exotic particles might have been created in large numbers in the very early Universe and contribute significantly to its overall mass density now but we would not be aware of them – the situation is similar to the problem of detecting the solar neutrinos, but now the flux of ultraweakly interacting particles would be very much lower than the flux of solar neutrinos. The challenge is to devise ways of finding out how much these different forms of matter contribute to the overall mass density of the Universe.

4.6 Gravitational lenses and dark matter

An ingenious way of searching for evidence of dark matter is to use the phenomenon of *gravitational lensing*. This phenomenon was predicted by Einstein on the basis of the general theory of relativity and was only observed in astronomical objects in 1979. We mentioned in Section 3.6 that the paths of light rays and all other types of electromagnetic radiation are bent by the gravitational influence of massive bodies. In the case of the Sun, light rays which just graze the visible edge of the Sun are deflected from a straight line by an angle of only 1.75 arcsec, which is a small angle, but which has now been measured with very high precision by radio interferometric measurements of compact radio sources. The gravitational deflection of light or radio waves by a massive object has the effect of focussing the light from a distant object, if the intervening object and the distant source are more or less in line. If the objects are precisely in line, the image of a background star would take the form of a circular ring about the lensing object. In general, the background source and the lensing galaxy will not be precisely lined up and then multiple images of the background object are expected (Figure 4.16(*a*))

The first gravitationally lensed quasar was discovered by Dennis Walsh, Robert Carswell and Ray Weymann in 1979 (Figure 4.16(*b*)). The two images of the quasar 0957+561 have identical spectra, as expected if they are different views of the same background object. One of the most intriguing aspects of this 'double quasar' is that the background quasar is variable and so the two images are also observed to vary but with a time delay associated with the different light travel times along the two paths illustrated in Figure 4.16(*a*). This a very important observation since it enables physical distances to be measured at the imaging galaxy, which are independent of knowledge of its redshift. This is one of the more promising new physical methods for measuring Hubble's constant.

In the 1980s, Bernard Burke and his colleagues undertook a major survey of a very large number of extragalactic radio sources to find other examples of gravitational lenses. They were successful in finding a number of candidates, among the most spectacular being the source MG1131 + 0456 which displays an almost perfect 'Einstein ring' as well as the double image of another part of the background source (Figure 4.17). It can be convincingly shown that this source is the gravitationally lensed image of a compact double radio source with a central radio component. One of the radio components is perfectly aligned with the lensing galaxy and the double image results from the central radio component being slightly misaligned.

The interest of these objects in the search for dark matter is that, if any dark object passes in front of a background source, it acts as a gravitational lens, resulting in the formation of multiple images, similar to the examples discussed above. Therefore, the number of gravitational lenses observed in a large sample of distant objects provides statistical evidence on the number of supermassive dark objects in the Universe. Jacqueline Hewitt and her colleagues have carried out such an analysis using the large survey of extragalactic radio sources described above. The relatively small number of gravitational lens candidates found enabled them to show that the mass density in supermassive black holes in intergalactic space must be less than the critical density.

Equally spectacular are the images of background objects gravitationally lensed by the cores of clusters of galaxies. Figure 4.18 is an image of the central region of the rich cluster of galaxies Abell 2218 imaged by the Hubble Space Telescope by Jean-Paul Kneib, Richard Ellis

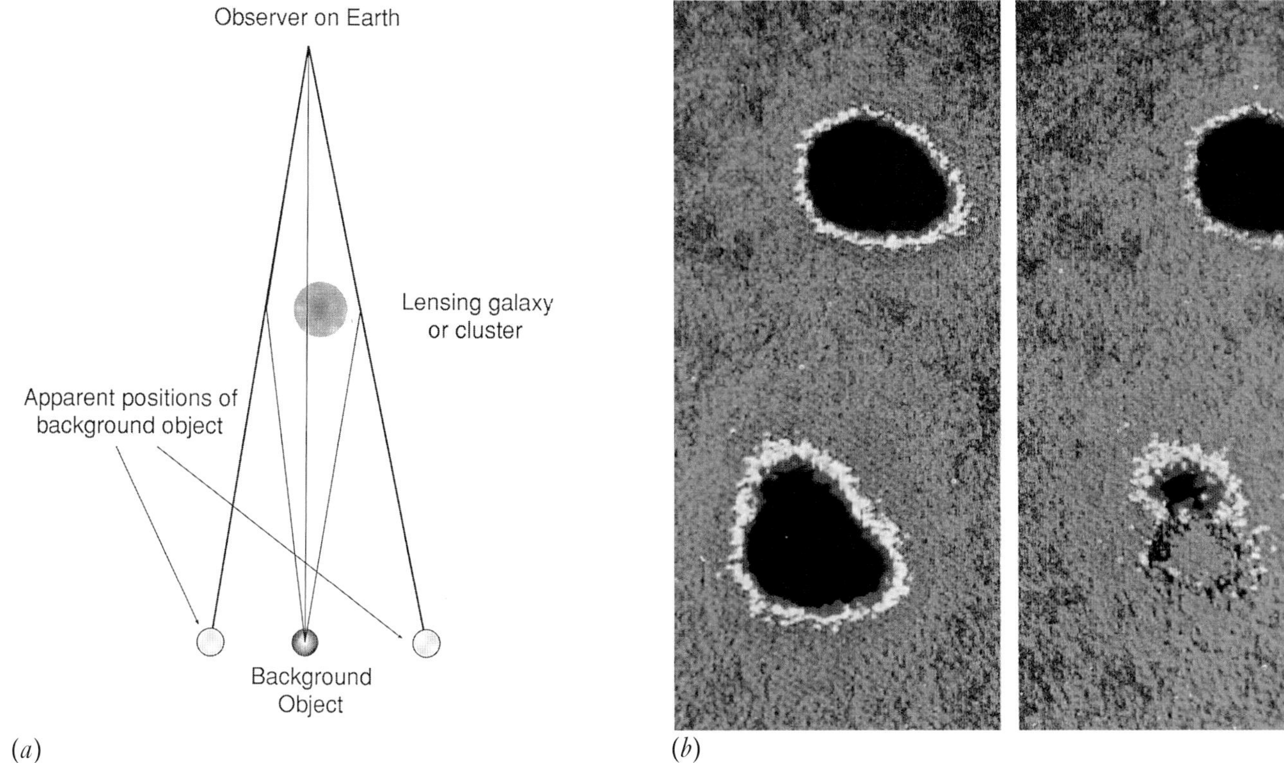

Figure 4.16. (*a*) Illustrating the formation of the gravitationally lensed images of a background source by an object lying along the line of sight. Two images of the distant object are formed because the light reaches the observer by two different routes. (*b*) An optical image of the double quasar 0957 + 561 A and B obtained by Alan Stockton. The images are in fact point-like and have been enhanced by image processing to show faint features. In the second picture, the lower image of the quasar has been subtracted, revealing the galaxy responsible for the lensing of the background quasar. The spectra of the two quasar images are identical.

and their colleagues. Remarkable narrow arcs are observed centred on the core of the cluster of galaxies and these are parts of Einstein rings associated with the imaging of very distant background objects. The properties of these arcs can be used to measure the total mass within the core of the cluster as well as providing information about the properties of the very distant lensed galaxy.

Another beautiful application of gravitational lensing to detect dark objects has concerned the nature of the dark halo about our own Galaxy. Just as in the case of the galaxy NGC 5084, our own Galaxy must possess a dark halo but the nature of the dark matter is uncertain. One possibility is that the halo consists of compact objects, perhaps brown dwarfs, 'Jupiters', old white dwarfs, neutron stars, or black holes, which would be very difficult to detect by the radiation they emit. These possible constituents of the halo of the Galaxy have been termed *massive compact halo objects*, or MACHOs. Occasionally, one of these dark objects passes in front of a background star and then the light of that star is gravitationally lensed. In the MACHO project, the objective is not to look for distorted images of the background stars, which would be too small to be observed, but to look for the brightening of the image of a background star as a MACHO passes in front of it. The expected brightening of the background star has a very specific variation of intensity with time and it should be independent of the wavelength of the radiation.

The MACHO project is a major effort to observe these gravitational lensing events. The most suitable background stars to observe are those belonging to our nearest neighbours in space, the Large and Small Magellanic Clouds. The concept behind the project is illustrated in Figure 4.19(*a*). The MACHOs must be in motion in the halo of our Galaxy and every now

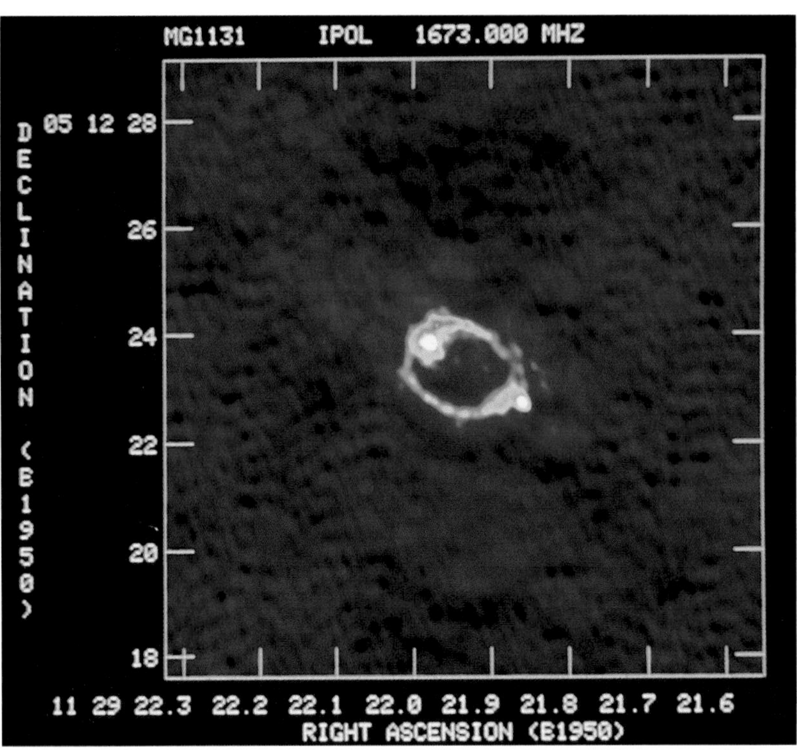

Figure 4.17. An image of the radio source MG1131+0456 made at 1.673 GHz with the MERLIN array, showing an almost perfect 'Einstein ring' due to the alignment of a component of the background radio source with the lensing galaxy. The ring is elliptical rather than circular because the lensing galaxy is ellipsoidal rather than spherically symmetric.

and then one of them passes directly across the line of sight between a star in the Magellanic Clouds and a telescope on Earth. There will then be a characteristic brightening of the star. In order to have a reasonable chance of detecting such events, it is necessary to monitor the brightnesses of over a million stars continuously and to be able to exclude known types of variable star. In 1993, the first report of gravitational brightening of a star in the Magellanic Clouds was reported (Figure 4.19(*b*)). The brightening of the star has exactly the correct time dependence and is independent of the wavelength at which the observations are made. The MACHO team estimate that the mass of the dark object responsible for the brightening must be between about 0.03 and 0.5 times the mass of the Sun, which would include a number of the possibilities mentioned above.

Similar results have been reported by the European EROS collaboration. These studies suggest that there are dark objects in the Galactic halo but the results of two years observations suggest that there are too few of these events for MACHOs to account for the dark matter in the Galactic halo. Current thinking is that the dark halo of our Galaxy may well consist of some form of exotic particle and detectors are now being built to search for the rare events when one of these ultraweakly interacting particles interacts within the sensitive volume of a detector placed far underground, away from contaminating background sources of penetrating radiation.

4.7 The basic problems of galaxy formation

At last, we can return to the problem of forming galaxies. The Friedman models are completely uniform and homogeneous and so they contain no structure at all. The hope is that, if there are tiny fluctuations in the density of matter from one point to another in the Universe, these grow under the influence of their own gravity. The Friedman world models provide

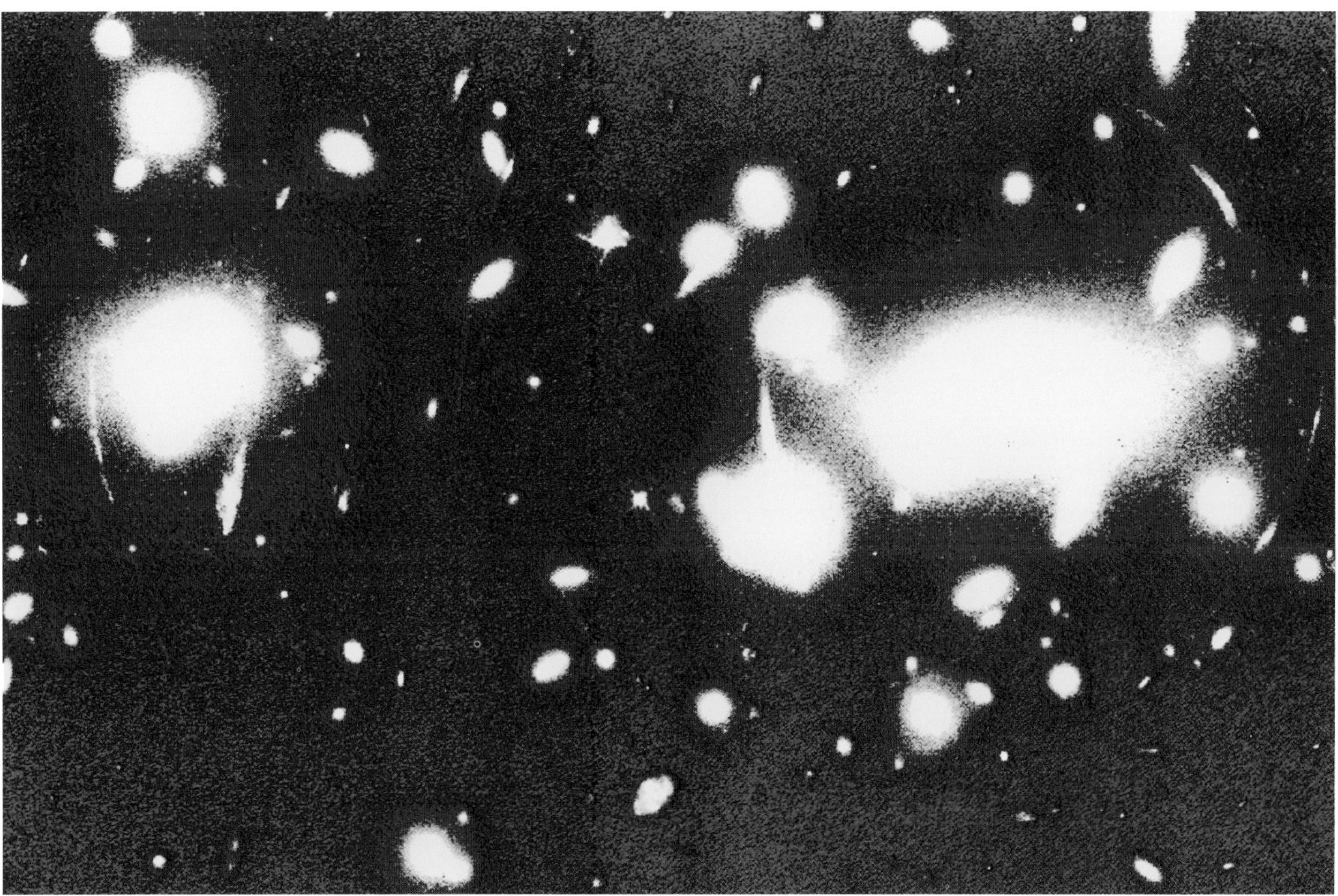

Figure 4.18. A deep image of the rich cluster of galaxies Abell 2218 observed with the Hubble Space Telescope by Jean-Paul Kneib, Richard Ellis and their colleagues. The narrow arcs centred on the core of the cluster of galaxies are partial Einstein rings associated with the gravitational lensing of distant background objects.

the background for studying this problem. The theorist poses the question in the following way, 'Suppose we begin with a completely uniform, expanding gas, the dynamics being described by the standard world models. Now, include tiny density fluctuations into the expanding gas. How do these density fluctuations develop as the Universe expands?' It might be hoped that the density fluctuations would increase exponentially in density, as is the case in a static medium, and which was described in Section 2.5. This is where the big problem arises – the density fluctuations do *not* grow exponentially in the expanding Universe.

It turns out that the density perturbations do grow but only very slowly. The reason is that, as soon as the density of any region begins to increase under the influence of its own gravity, most the increase is undone, because the gas is expanding as a whole and thus decreases its average density. As a result, the gas feels a reduced gravitational pull as compared with the static case, and the density increases very slowly. We can illustrate the origin of this problem by carving out a spherical region of the Universe and squashing it to a slightly higher density than the background (Figure 4.20(*a*)). We can then work out how the density of that region would grow relative to the background density by treating the little region as a closed Universe of slightly higher density than the background. This is exactly the same calculation as was carried out in Section 4.3 for the Universe as a whole and which resulted in the

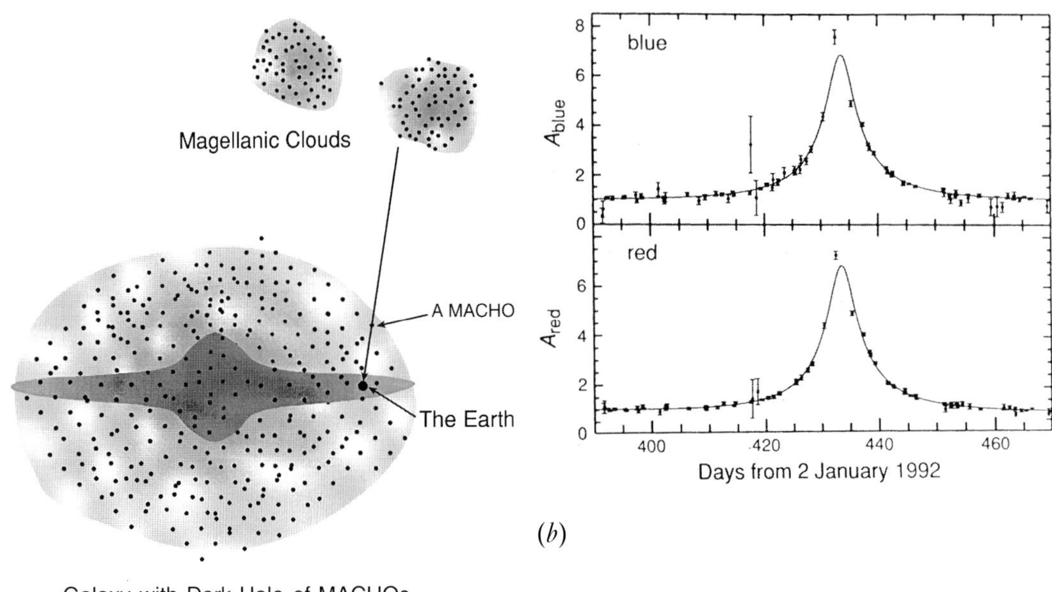

Figure 4.19. (*a*) Illustrating how objects in the halo of our Galaxy can lead to the gravitational lensing of background stars in the Magellanic Clouds. (*b*) A gravitational lensing event recorded by the MACHO project in February and March 1993. The horizontal axis shows the date in days as measured from day zero which was 2 January 1992. The vertical axes show the amplification of the brightness of the lensed star relative to its unlensed intensity in standard blue and red wavebands. The solid lines show the expected variation of the brightness of the lensed object with time. It has the same characteristic shape at red and blue wavelengths.

dynamics of the world models illustrated in Figure 4.9. This model provides an exact description of the growth of a spherically symmetric density fluctuation relative to the background density.

The key result concerns the rate at which density perturbations grow as the Universe expands. If the background density of the Universe is ρ and the excess density in the region is $\partial\rho$, so that its total density is $\rho+\partial\rho$, the quantity of interest is the rate at which the density enhancement $\partial\rho$ grows relative to the background density ρ. The quantity $\partial\rho/\rho$ is often referred to as the *density contrast* of the perturbation. If the Universe is of high enough density, the density contrast grows as

$$\frac{\delta\rho}{\rho} \propto R$$

where R is the scale factor of the Universe, the quantity which describes how the distance between any two typical points in the expanding Universe separate with time. The key point is that the growth of the density contrast is *not* exponential, but only *linear* in the scale factor. We can understand why the growth of the perturbation is not exponential by considering the development of the spherical density perturbation relative to the background, as illustrated in Figure 4.20(*a*). Let us suppose that the background has the critical density, $\Omega = 1$. Then, the dynamics of the spherical region must correspond to a Universe with slightly greater density, which will eventually collapse to form some sort of bound object as illustrated in Figure 4.20(*b*). Notice that the line representing the model with $\Omega = 1$ and that corresponding to slightly greater density diverge very slowly indeed and certainly not exponentially.

We can learn more from the Figure 4.20(*b*). It can be seen that, if the background model had density less than the critical density, then a tiny density perturbation might not be sufficient to increase the density of that region to such a value that it would lie on a trajectory which results in collapse, that is, the region of enhanced density would not have $\Omega > 1$. It is

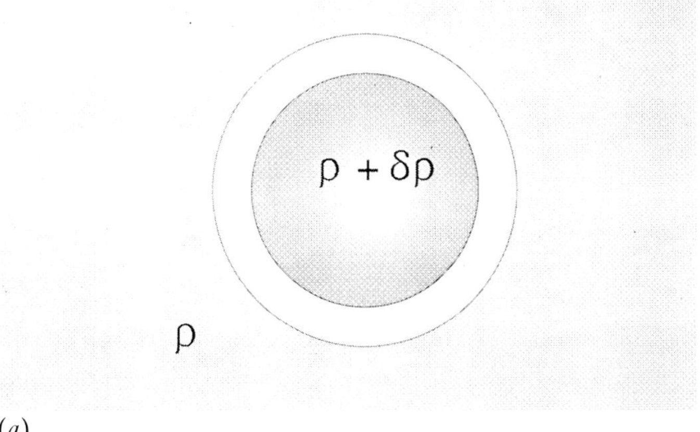

(*a*)

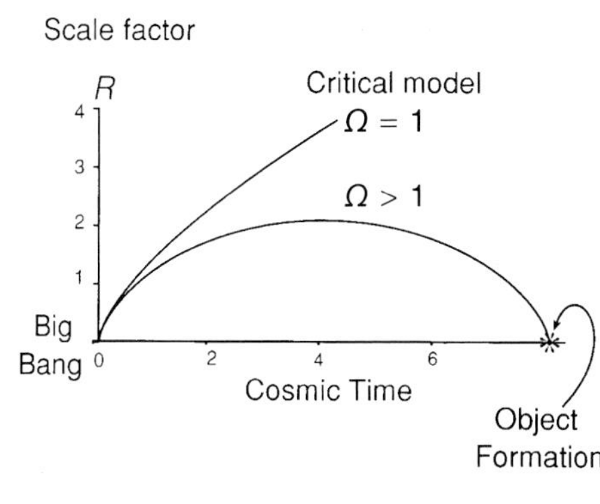

(*b*)

Figure 4.20. (*a*) Illustrating a spherical region of enhanced density $\rho + \partial\rho$, embedded in a background of density ρ. The region behaves like an expanding universe of slightly enhanced density relative to the background model. (*b*) Illustrating the dynamics of the enhanced region relative to the background model. It can be seen that the enhanced region behaves exactly like a world model with density parameter $\Omega > 1$ which eventually collapses to a high density.

a general result that the growth of the perturbation to form collapsed objects, such as galaxies and clusters of galaxies, only takes place according to the above relation if $\Omega/R > 1$.

These are the crucial results we have been seeking and they lead to a profound problem. We might have hoped that galaxies would have formed from the gravitational collapse of regions of slightly enhanced density in the early Universe. Tiny density fluctuations must occur statistically but they are only useful if the fluctuations are able to grow exponentially under gravity, so that they can grow to very large densities in a finite time. This is the attractive feature of exponential growth. Linear growth, of the type described above, is far too slow to make finite size perturbations from infinitesimal perturbations in the early Universe.

Theorists have tried very hard to find a way round this problem but no solution has been found. It was first recognised to be a problem by Lemaître and Tolman in the 1930s and the complete solution was found by Lifshitz in 1946. These authors regarded this as such a fundamental problem for galaxy formation that they believed that galaxies and clusters of galaxies could not have formed by gravitational collapse. The gravitational instability grows so slowly that finite sized density enhancements would have to be included into the initial conditions from which the Universe expanded and this did not seem a very attractive proposition.

Subsequent generations of theorists have, however, adopted a different point of view and assumed that, for one reason or another, finite sized density perturbations were present in the early Universe and that the structures we observe today formed from these by the type of gravitational instability described above. Furthermore, it is possible to search for traces of the early development of these fluctuations as a result of the discovery of the cosmic microwave background radiation. To understand how this comes about, we have to study the thermal history of the Universe.

4.8 Radiation in the expanding Universe

From the considerations of Section 4.3, we know how the scale factor R changes with cosmic epoch, and so how the matter density changes, since the density of matter at any earlier epoch is $1/R^3$ times its value at the present epoch. Now, we have to consider what happens to the radiation as well. The cosmic microwave background radiation is remarkably isotropic and has a perfect black body spectrum at a temperature of 2.725 K. The radiation has a remarkably large energy density at the present day. If we use Einstein's mass–energy relation,

$E = mc^2$, the mass density in this radiation at the present day corresponds to about one ten thousandth of the critical density. We will argue that this radiation originated in the Big Bang, so what happens when the radiation as well as the matter is squashed backwards in time?

There are two beautiful results which give us a simple picture of how the properties of electromagnetic radiation change as the Universe expands. The first of these concerns the real meaning of redshift in cosmology. In Section 4.2, we described how Hubble discovered that the distribution of galaxies is expanding uniformly. He measured the recessional velocities of the galaxies by measuring their *redshifts*, that is, the shift of the spectrum of a galaxy to longer wavelengths, because of its motion away from our Galaxy. This is the Doppler shift of the radiation. We define the *redshift*, z, of a signal by the formula

$$z = \frac{\lambda_{obs} - \lambda_{em}}{\lambda_{em}}$$

where λ_{em} is the wavelength with which the radiation is emitted and λ_{obs} is the observed wavelength, which is greater than λ_{em} because of the motion of the source. If the velocity of the source, V, is not too great, meaning that the velocity does not approach the speed of light, there is a simple relation between the redshift and the velocity of the source

$$V = cz$$

where c is the speed of light. Redshifts of galaxies can be measured from the displacement of the lines in their spectra from their rest wavelengths. Recessional velocities can then be found by multiplying the redshifts by the speed of light.

In fact, the redshift has a much deeper meaning in cosmology. The greater the distance of a galaxy from our own Galaxy, the greater its velocity of recession, and also the longer the light took to reach us – at the time when the light was emitted, the average distance between galaxies was smaller than it is now by the scale factor R, which describes the dynamics of the expansion of the Universe. It turns out that the redshift is simply related to the scale factor R. The key result is:

$$R = \frac{1}{1+z}.$$

What this equation means is that, if we measure the redshift of a galaxy or quasar, we know immediately the value of R when the light was emitted. For example, a galaxy with redshift 1 emitted the light we observe today when the scale factor of the Universe was only half its present value, that is, all the galaxies were on average twice as close together as they are today. Quasars with redshift 4 emitted their light when the scale factor of the Universe was only one fifth of its present value, and so on. Thus, we know precisely the value of the scale factor when the light was emitted but we do not know at what cosmic time this occurred – we have to use the cosmological models to determine the relation between R and cosmic time.

We can reorganise the relation between the scale factor and redshift so that the relations between the emitted and observed frequencies or wavelengths are expressed in terms of R. We find

$$\frac{\nu_{obs}}{\nu_{em}} = R. \qquad \frac{\lambda_{em}}{\lambda_{obs}} = R.$$

The second relation tells us how the wavelength of radiation changes in the expanding Universe. It can be seen that the wavelength increases exactly as the average distance between galaxies increases. In particular, we can now work out how the spectrum of the cosmic microwave background radiation changes as the Universe expands. The frequency of the waves increases as $1/R$ as the Universe is squashed backwards in time and so the maximum of the black body spectrum moves to higher temperature, as described in Section 1.3. In fact, the temperature of the radiation increases inversely proportional to R as we squash the radiation back in time. Thus, at a redshift of 1, the scale factor was $\frac{1}{2}$ and the radiation temperature of the background radiation was $T_{\mathrm{rad}} = 2.725 \times 2 = 5.45\mathrm{K}$. This is the key relation we need to study the thermal history of the Universe, namely,

$$T_{\mathrm{rad}} \propto \frac{1}{R}.$$

We will study the full implications of this important formula in Chapter 5 but, for the moment, we need only study the relatively recent thermal history of the Universe.

As we squash the Universe back in time, the temperature of the cosmic microwave background radiation increases, as described above. One of the most important epochs occurred when the scale factor had a value of about 1/1500, about 300 000 years after the Big Bang (Figure 4.21). At that point, the radiation temperature of the background radiation was $T = 1500 \times 2.725\ \mathrm{K} \approx 4000\ \mathrm{K}$. Now, hydrogen is by far the most abundant element in the Universe. At this temperature, all the hydrogen in the Universe is broken up into its constituent protons and electrons by the ultraviolet radiation in the high energy tail of the spectrum of the cosmic microwave background radiation, which, at a temperature of 4000 K, is now glowing like the surface of a star just slightly cooler than the Sun. The breaking up of hydrogen into its constituent protons and electrons is known as *ionisation* and this has important consequences for the interaction of radiation and matter in the Universe. The most important of these is that matter in the form of free electrons and protons is what is known as a *plasma* and it has quite different electrical properties from neutral hydrogen. In particular, the coupling between the radiation and the matter becomes very strong indeed. The negatively charged electrons scatter the background radiation very effectively and therefore we can think of the matter and radiation as being tightly coupled together by these scattering processes. This is similar to what happens when we observe the surface of the Sun. The matter in the interior of the Sun is ionised and so radiation is very strongly scattered as it tries to escape from the Sun. In fact, we only observe the very surface layers of the Sun from which the radiation was last scattered. Because the scattering is so strong, we cannot obtain directly information from inside the Sun – we only observe radiation from what is known as a *last scattering surface*.

Exactly the same process takes place when we look further and further back into the early Universe. When the Universe had scale factor $R \approx 1/1500$, all the hydrogen in the Universe was fully ionised, and, at all earlier epochs, the pregalactic hydrogen was in the form of a fully ionised plasma. Thus, when the clocks are run forward, the primordial ionised hydrogen recombines to form neutral hydrogen and this key epoch in the history of the Universe is therefore known as the *epoch of recombination*. As a result of this strong scattering of the radiation by the free electrons, detailed calculations show that we cannot observe any electromagnetic radiation originating from redshifts $z > 1000$. It is just as if we were observing

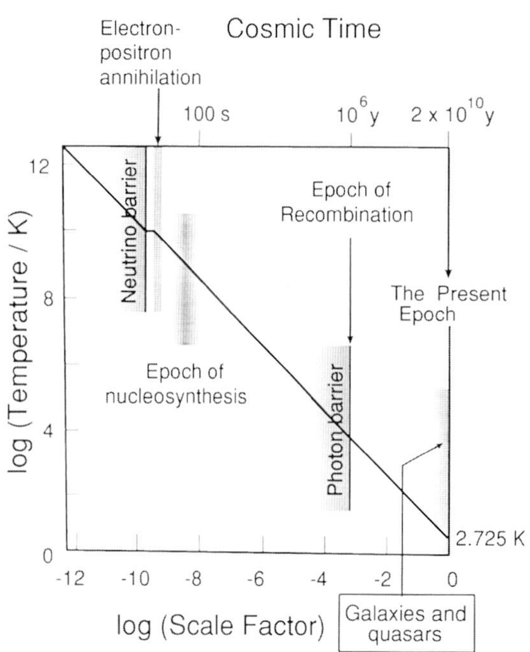

Figure 4.21. Illustrating the temperature history of the Universe in the recent past. The solid line indicates how the temperature of the cosmic background radiation has changed with scale factor *R*. The diagram also shows certain key epochs. The epochs over which galaxies and quasars can be observed are indicated, as well as the 'epoch of recombination', prior to which all the hydrogen in the Universe were fully ionised. At early epochs, the dynamics of the Universe were dominated by the inertial mass density of the cosmic background radiation – these epochs are referred to as the 'radiation-dominated' era.

the Sun all around us. The ionised gas at the redshift $z \approx 1000$ is the *last scattering surface* for the cosmic microwave background radiation. We do not observe radiation from this last scattering surface at the temperature of 4000 K because it has been redshifted to its present temperature of 2.725 K from a redshift of 1000 to the present epoch.

It can now be appreciated why this result is so important for studying the origin of galaxies and other large scale structures in the Universe. We can work out how large the perturbations in density have to be in the past in order to ensure that galaxies and clusters of galaxies are formed by the present epoch. We know that galaxies exist at the present epoch and so, at the very least, they must have achieved a density contrast of $\partial\rho/\rho \sim 1$, and preferably very much greater values, by now. Now, the fastest growth rate for structure in the Universe occurs in the critical $\Omega = 1$ model, in which the density contrast grows proportional to R. Therefore, we can work out the minimum density contrast which must have been present in the matter at the last scattering surface at $z \approx 1000$. It follows that the density contrast must have been at least $\partial\rho/\rho \approx 1/1000$ on this last scattering surface. Because the matter was strongly coupled to the radiation, if there were fluctuations in the matter distribution, we expect to observe similar perturbations in the radiation as well. Detailed calculations show that we would expect to observe fluctuations in the intensity of the cosmic microwave background radiation over the sky at a level of about one part in 1000 due to the formation of large scale structures such as clusters of galaxies. This is in clear contradiction with the extreme isotropy of the cosmic microwave background radiation, in which fluctuations have only been observed at a level of about one part in 100 000.

This is a devastating blow for this simplest of pictures of how galaxies might have formed in the expanding Universe. If there were only ordinary matter present in the Universe, it is very difficult to avoid the conclusion that we ought already to have seen traces of the formation of galaxies and larger scale structures in the distribution of the cosmic microwave background radiation over the sky. There is, however, a solution at hand and that is that, so far, we have only considered the behaviour of ordinary matter of the Universe. We showed in Section 4.5 that most of the mass in the Universe is in the form of dark matter and this need not necessarily behave like the ordinary forms of matter at all. Let us see how the dark matter helps.

4.9 Dark matter and galaxy formation

The problem to be solved is how to ensure that large scale structures such as galaxies, clusters of galaxies and the large voids are formed by the present day, whilst at the same time the fluctuations in the cosmic microwave background radiation are reduced to the observed low level of only about one part in 100 000. The solution favoured by most cosmologists results from the realisation that most of the matter in the Universe cannot be in the form of ordinary matter – rather, as we will discuss in more detail in Section 5.3, most of the dark matter has to be in some exotic form and the massive neutrinos and the various possible forms of weakly interacting massive particles, or WIMPs, have attracted particular attention. These types of particle are predicted by current theories of elementary particles but they have not yet been detected in experiments with large accelerators, since the energies at which they would be created far exceed those which these machines can provide. In other words, these hypothetical particles are based upon extrapolations of current theories of elementary particles far beyond the energies at which they have been confirmed by laboratory experiments.

These forms of dark matter interact very weakly indeed with ordinary matter and, in the cosmological context, they only interact with ordinary matter through their gravitational influence. Suppose that there is much more matter in the Universe in the form of dark matter than there is in the form of ordinary matter. The clue to understanding what happens is to remember that, when the scale factor of the Universe was less than about 1/1000, the ordinary matter was very strongly coupled to the radiation and this made it very difficult indeed for the ordinary matter to collapse to form galaxies and clusters of galaxies prior to this epoch. The coupling was so strong that the pressure of the radiation within the perturbations prevented the matter from collapsing to form condensed objects through the Jeans instability. The consequence was that during the pre-recombination eras, the perturbations in the ordinary matter were stabilised and they behaved like sound waves in the plasma.

The dark matter did not, however, interact with the ordinary matter or the radiation at all during these early epochs. Therefore, density perturbations in the dark matter could grow perfectly happily according to the relation we discussed in Section 4.7, while the ordinary matter was prevented from collapsing because of its strong coupling to the radiation prior to the epoch when the hydrogen recombined. Thus, we can imagine perturbations in the dark matter having developed to quite large density contrasts by the epoch of recombination, while the perturbations in the ordinary matter had hardly grown at all because they were strongly coupled to the radiation. When the hydrogen recombined at a redshift of about 1000, however, the ordinary matter suddenly became neutral and was no longer coupled to the radiation. It was then free to collapse into perturbations already present in the dark matter and the perturbations in the ordinary matter grew very rapidly to the same density contrast as those in the dark matter. The attraction of this picture is that, at the time when the temperature fluctuations were imprinted on the cosmic microwave background radiation, the fluctuations in the ordinary matter were very much smaller than those in the dark matter and, in particular, they were much smaller than the perturbations in the simple model described in Section 4.7. This is how it is possible to account for the very small fluctuations in the matter, which was coupled to the background radiation, while the fluctuations in the dark matter were sufficiently large to ensure that galaxies and larger scale structures formed by the present epoch.

This is the currently favoured picture for the origin of the large scale structure of the Universe and it has been the subject of a great deal of detailed theoretical study. There are two preferred versions of the dark matter theories. In one version, called the *cold dark matter* picture, the dark matter consists of weakly interacting massive particles, the WIMPs, or their close relatives. This form of dark matter is called 'cold' because the particles have to be massive or else they would already have been detected in high energy particle experiments. One of the most important results of the Large Electron–Positron (LEP) collider experiments at CERN is that some of the possible forms of dark matter can already be excluded. For example, it has been shown that there are only three types of neutrino. This result came from studies of the decay modes of the particles known as the $W^{\pm}$ and Z^0 bosons, which are the particles involved in transmitting the electroweak force. These particles were discovered in high energy LEP experiments. If the $W^{\pm}$ and Z^0 bosons were able to decay into any of these exotic particles, their presence would have already been detected in the LEP experiment, if their rest mass energies were less than about 40 GeV. This rest mass energy corresponds to about 40 times that of the proton. Consequently, if these ultraweakly interacting particles exist, they must be massive. As a result, they could only have come into equilibrium with other forms of matter in the very early Universe and so they must be very cold at the present epoch. In this scenario, primordial density fluctuations survive on all scales from the very early Universe. The initial spectrum of the fluctuations contains objects of a very wide range of masses and, after the epoch when the primordial plasma recombines, the ordinary matter falls into the perturbations in the dark matter and begins the process of forming distinct objects. We can think of the ordinary matter as tracers for the distribution of the dark matter perturbations. The low-mass objects begin to cluster and coalesce to form more massive objects and, in due course, these objects form yet more massive objects. This process of galaxy formation is known as *hierarchical clustering*. It can be thought of as a 'bottom-up' process in which the largest scale structures form latest in the Universe and are built up out of smaller mass objects.

An alternative version of the dark matter picture is known as the *hot dark matter* theory. In its simplest form, it is supposed that the three types of neutrino have a finite rest mass of about 10 eV. This is a very interesting mass because, according to the standard Big Bang model of the Universe, the number of primordial neutrinos is known very precisely and, if they had this mass, the mass density of the neutrinos would be sufficient to close the Universe, that is, their mass density would correspond to $\Omega = 1$. This picture of galaxy formation is rather different from the cold dark matter picture because all small scale structure is wiped out long before the epoch at which the primordial plasma recombines – the neutrinos interact very weakly indeed with matter and so can stream freely out of the perturbations. In this picture, only the very largest scale structures survive to the epoch of recombination and these are the structures which eventually collapse to form large scale structures in the Universe. The clusters of galaxies, galaxies and smaller scale systems form by the sequential fragmentation of these large scale structures. This is sometimes referred to as a 'top-down' process, in which the largest scale structures form first.

It is intriguing that the two simplest dark matter models for the formation of structure in the Universe result in two completely different pictures of how galaxies were formed. In the cold dark matter picture, the galaxies are built up from smaller building blocks – in the hot dark matter picture, the galaxies form by the fragmentation of larger scale structures. In the former, galaxy formation begins soon after the epoch of recombination, whereas, in the latter, galaxy formation begins late in the Universe.

In all models, some input spectrum of the initial density fluctuations has to be assumed and a number of remarkable computer simulations have been made of the way in which the small perturbations in the dark matter develop under gravity. Figure 4.22 shows the results of computer simulations of the expected large scale structure of the Universe in the two different versions of the dark matter picture for the origin of structure in the Universe. It can be seen that the models are reasonably successful in accounting for large scale structures. There are, however, problems with both of the models. In the case of the cold dark matter model, it is difficult to account for the enormous voids, walls and stringy structures seen in the Harvard Center for Astrophysics survey (Figure 1.17). This can be understood from the physical point of view because the process of formation of structure is hierarchical clustering which tends to wipe out stringy structures. On the other hand, the hot dark matter picture produces far too much large scale structure – this is because the matter cools on falling into

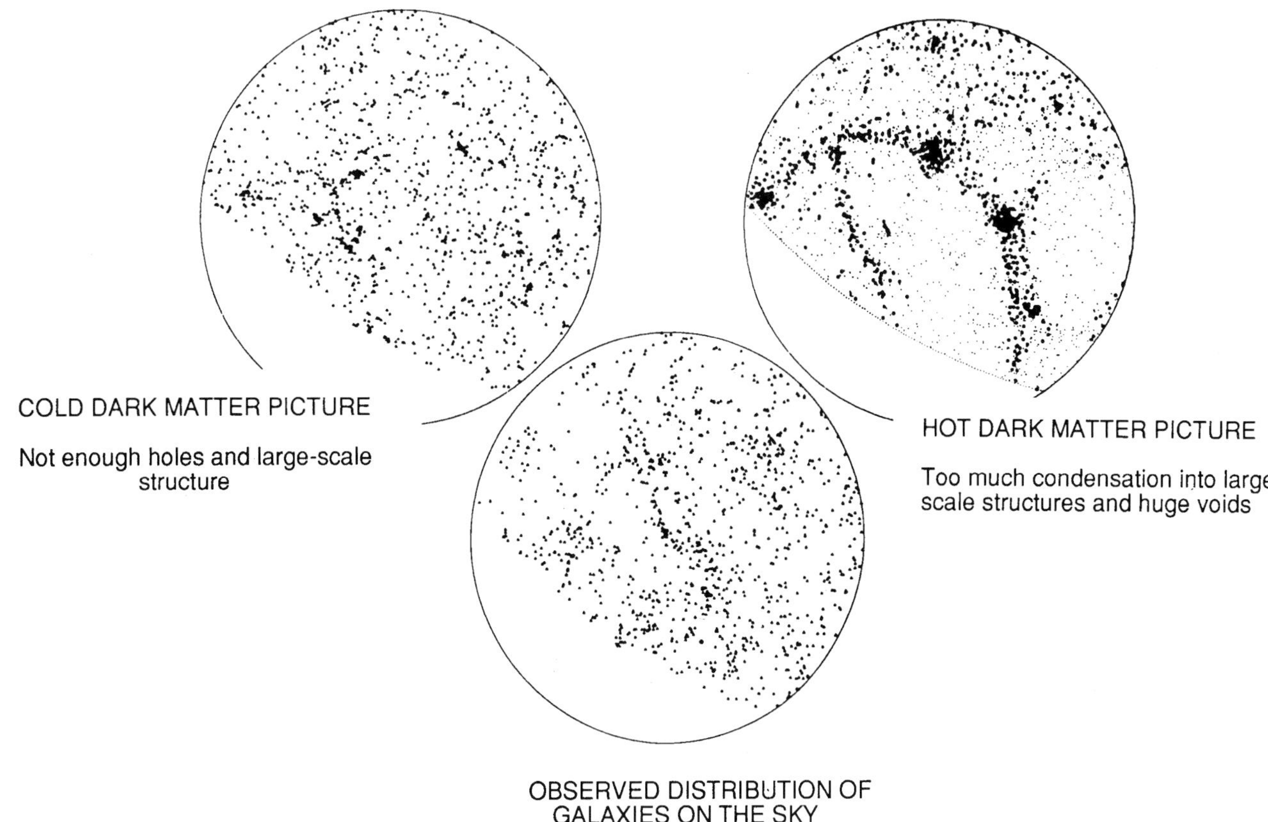

Figure 4.22. Simulations by Carlos Frenk of the expected large scale distribution of galaxies according to the cold dark matter model and the hot dark matter model of the origin of the large scale structure of the Universe. These predictions are compared with the observations. The simple cold dark matter model does not produce sufficient large scale structure in the form of voids and filaments of galaxies, whereas the simple hot dark matter model produces too much large scale structure.

the giant gravitational potential wells created by the dark matter and all the galaxies form in thin collapsed sheets. Clearly, the existing models are only partially successful in accounting for the observed distribution of matter in the Universe.

The current consensus of opinion favours the cold dark matter picture, which has a number of attractive features. It can account for the observed clustering of galaxies over a very wide range of physical scales, whereas the hot dark matter picture results in too much structure on large scales. A key test of the cold dark matter picture is whether or not it is consistent with the intensity fluctuations in the cosmic microwave background radiation observed by the COBE satellite. The millimetre map of the sky observed by the COBE satellite (Figure 4.6) was made with an angular resolution of 10°. As described in Section 4.8, the cosmic microwave background radiation we observe today originated on the last scattering surface, which occurred at a redshift $z \approx 1000$. We can work out the physical scale in the Universe today which corresponds to 10° on the last scattering surface. This size turns out to be about ten times the size of the large holes seen in Figure 1.17, in other words, these fluctuations are on scales very much greater than those of galaxies, clusters of galaxies and even the large voids. Thus, these fluctuations cannot be directly compared with the structures observed in the Universe today. It is, however, reasonable to extrapolate the initial spectrum of the density fluctuations in the cold dark matter picture, which can successfully account for clustering of galaxies on scales up to those of clusters of galaxies and greater, to the scale of the COBE fluctuation.

What is the origin of the fluctuations in the cosmic microwave background radiation observed by COBE? Density fluctuations of different masses at the last scattering surface are illustrated in Figure 4.23. As can be seen from the diagram, the last scattering surface from which the radiation reaches us is not an abrupt transition but there is a range of redshifts from within which the radiation was last scattered. Detailed calculations show that 50% of the radiation we observe at the present day was last scattered between redshifts 1130 and 1010. The sizes of different large scale structures relative to the width of this last scattering layer are illustrated in Figure 4.23. It can be seen that the size of perturbations on the scale of the COBE beam, 10°, is very much greater than the thickness of the last scattering layer. The intensity fluctuations in the background radiation are due to the fact that the radiation originating within the last scattering surface has to 'climb out' of these density fluctuations. This phenomenon is known as the *gravitational redshift* of the radiation and results in a slight decrease in the intensity of the radiation in the direction of the fluctuation. On the scale of the giant holes, the fluctuations have scale roughly ten times the thickness of the last scattering layer. If the spectrum

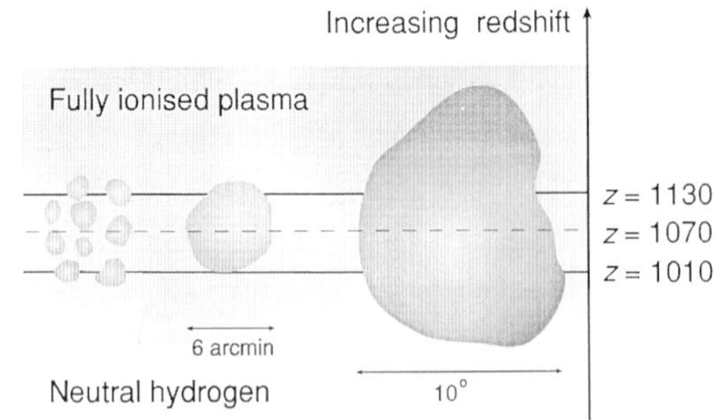

Figure 4.23. Illustrating the origin of intensity fluctuations in the cosmic microwave background radiation. The recombination of the primordial plasma does not take place instantaneously but takes place over a range of redshifts. The lines labelled $z = 1130$ and $z = 1010$ are the redshifts between which 50% of the radiation arriving at the Earth was last scattered. The blobs represent fluctuations of different masses and their sizes can be compared with the thickness of the last scattering layer.

of density fluctuations which can account for the large scale distribution of galaxies is adopted, it turns out that intensity fluctuations of amplitude of about one part in 100 000 are expected, in good agreement with the fluctuations observed in the cosmic microwave background radiation. These are very encouraging results and suggest that this picture of the origin of the large scale structure of the Universe is probably along the right lines.

Despite these successes, there remains the fundamental problem, identified by Lemaître, Tolman and Lifshitz, that an initial perturbation spectrum with finite amplitude is essential if we are to make the galaxies, clusters of galaxies and larger scale structures we observe in the Universe today. This is very worrisome because it means that we only get out of our model what we put in at the beginning. This is the first of four features which, according to the classical theory, have to be built into the initial conditions of our Universe if we are to create the structures seen today. We will encounter three other problems of this sort in Chapter 5.

4.10 Making real galaxies

It may come as a surprise that we have spent so much effort trying to disentangle the very earliest stages of the formation of galaxies, in particular, in trying to understand how the gas out of which the galaxies first formed collapsed. Ideally, we would like to make observations of very distant objects and simply read off directly the sequence of events which resulted in the creation of galaxies as we know them today. In other words, simply look further and further back along our past light cone (Figure 1.18) and take images of the Universe at earlier and earlier cosmic epochs. The big problem is that normal galaxies, such as our own Galaxy, cannot be observed at large redshifts because they are not luminous enough to be detectable. The most distant samples of normal galaxies which can be observed with the present generation of large telescopes, extend only to redshifts of about 0.5 to 1, meaning that we can only study the evolution of galaxies from the time the Universe was about half its present age to the present day.

Ultraluminous objects such as quasars and certain radio galaxies can be observed back to redshifts of about 4, when the scale factor of the Universe was only about one fifth of its present value. This means that we can study the properties of the most luminous objects back to times when the Universe was probably only about a tenth of its present age. These are, however, very rare and special objects and they do not tell us about the ordinary stuff of the Universe. It is possible to use them as background sources and study intervening objects by the absorption features which they imprint on the spectrum of the distant object and this is an important way of studying intergalactic clouds which may be the precursors of galaxies. The quasars are the most distant objects we can observe before we encounter the last scattering surface of the cosmic microwave background radiation at a redshift of about 1000. Between redshifts 4 and 1000, there is a great gap, which is very difficult to study observationally, and yet it was precisely during these epochs that much of the activity which resulted in the formation of the large scale structure of the Universe as we know it must have occurred.

What have we learned about the behaviour of galaxies during the more recent, accessible epochs? The most extensive work has been carried out on very active galaxies, in particular, the radio galaxies and the quasars. These are the most luminous objects in the Universe and can therefore be observed at much greater distances than ordinary galaxies. When large

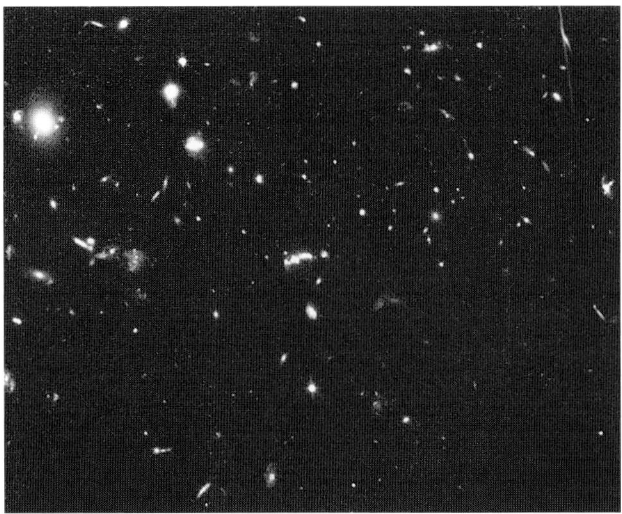

Figure 4.24. The field of the radio galaxy 3C 324, which is a member of a cluster of galaxies, observed by the Hubble Space Telescope by Marc Dickinson and his colleagues. The radio galaxy and the cluster have redshift $z = 1.12$. The radio galaxy has a strange appearance and seems to consist of a number of interacting components. There are many strongly interacting galaxies in the cluster as well as a number of elliptical galaxies.

samples of distant radio galaxies and quasars are surveyed, it is found that they were very much more common in the past than they are at the present epoch. It seems that, when the Universe was about a quarter or a fifth of its present age, the peak of quasar and radio galaxy activity must have occurred. This means that the process of galaxy formation must have been such that these galaxies had time to grow very massive black holes in their nuclei, similar to those observed in M87, by the time the Universe was only about a fifth of its present age. This is an important constraint upon the formation and evolution of these galaxies. Some of the most distant radio galaxies are associated with huge clouds of ionised gas. An intriguing question is what the evolutionary status of these clouds is. Are these the remnants of the gas clouds out of which the galaxy formed or are they simply gas clouds in the vicinity of the galaxy, which are excited by the activity going on in the active nucleus? We do not know the answer to these questions, nor do we understand the role of such events in the scheme of galactic evolution. The problem with using quasars and radio galaxies in these studies is that they are rather exceptional objects and the astrophysical processes responsible for the emissions we observe are much more poorly understood than those of stars and galaxies.

Despite these problems, tools are becoming available to address the questions of the origin and evolution of all types of galaxy. The Hubble Space Telescope is now producing magnificent images of galaxies at redshifts of about 1. Figure 4.24 shows a recent image of a cluster of galaxies at a redshift of 1 associated with the radio source 3C 324. There are many remarkable features in this image, including the structure of the radio galaxy which bears little resemblance to a giant elliptical galaxy. It can be seen that there are many disturbed and interacting systems in the cluster in addition to a number of objects which appear to be elliptical galaxies. In general terms, this image indicates that the cluster is in a dynamically young state. It is very encouraging that such images of galaxies can now be obtained by the Hubble Space Telescope. The significance of such observations is that we can now begin the detailed study of galaxies when the Universe was only about half its present age. The next generation of very large ground-based optical-infrared telescopes is very important for following up these types of observation. Taking an image of a distant galaxy provides a great deal of information but, in order to understand the stellar content and evolutionary status of these distant

Figure 4.25. A photograph of a model of one of the 8-metre Gemini telescopes. The large telescopes of the future are generally thin-mirror telescopes and the figure of the mirror is maintained precisely by computer control.

galaxies, it is essential to obtain their spectra and this requires the full power of the next generation of 8–10 metre optical-infrared telescopes, such as the Keck and Gemini telescopes (Figure 4.25).

The exciting thing from the point of view of the future of cosmology is that we can now observe galaxies and active galaxies when the Universe was significantly younger than it is now. We can therefore begin putting the subject of the origin and evolution of galaxies, their environments and the large scale structure of the Universe on a firm observational footing rather than this subject simply being the province of theoretical speculation. These studies have very profound implications for cosmology because the number of real facts which we have about the evolution of the Universe is still quite limited. The more we can pin down directly by observation about how the galaxies and the large scale structure have evolved into their present state, the more secure our understanding of the evolution of the Universe will be. It may be that we will eventually be able to determine, directly by observation, how the galaxies and the large scale structure of the Universe came about. This is a challenging programme but at least we know exactly what we would like to do observationally.

5 The origin of the Universe

5.1 Towards the Big Bang

Finally, we have to grapple with the biggest question of them all – the origin of the Universe itself. In the last chapter, we found that there are features of the Universe which, according to classical physics, must be built into the initial conditions from which our Universe evolved. Specifically, we showed that there must be density fluctuations in the very early Universe to ensure that galaxies, clusters of galaxies and larger scale structures were formed by the present day. In this chapter, we are to study three further problems, again within the context of the Big Bang model of the Universe, which force us to endow the very early Universe with quite specific properties, if we are to reproduce the Universe we observe today. Before embarking on this journey backwards in time towards the origin of the Big Bang, let us consider some words of William McCrea, which I believe are very important for the scientific study of cosmology.

In 1970, McCrea wrote a paper entitled 'A Philosophy for Big-Bang Cosmology', in which he asked how much we can hope to learn about the very early stages of the Universe from observations we make today. As he pointed out, there is a fundamental problem at the outset. We have only one Universe to study and that distinguishes the scientific study of cosmology from all other sciences. In physics, critical experiments can be carried out by independent workers with completely different sets of apparatus and it is the agreement and repeatability of these experiments which gives us confidence in the results of the experiments and their implications for theory. In the case of the Universe, we have only one example and we cannot even do experiments with it. All we can do is observe it. We can, to some extent, carry out independent experiments by repeating observations in different regions of the Universe and, if the same results are found wherever we look, we can suspect that we have found a general rule. These results are, however, applicable only for the region of the Universe we can observe and there might be some properties of the Universe which are only observable on the very largest scales, for which there is no possibility of making an independent observation. Thus, for certain global aspects of our Universe, we cannot obtain any independent check of our observations.

Of the six propositions in McCrea's paper, two are of special significance for current cosmological research.

> **Proposition (B)** *The less information we can get, the less we need in order to make predictions that are confirmed by observation.*
>
> **Proposition (C)** *From the observed properties of the present state of the Universe, we can infer less and less about earlier and earlier previous states, and almost nothing about what we might wish to call the initial state.*

It is worthwhile pondering the full significance of these remarks, especially in the light of some of the remarkable claims which have been made recently about our understanding of the early stages of our Universe. If these earliest phases are more or less inaccessible to observational study, there is considerable freedom in the choice of conceivable physical theories.

We may be guided by our knowledge of laboratory physics, but how can we test that these are the correct laws to use in the extreme conditions of the very early Universe? We therefore need to ask how valid these views are, particularly Proposition (C), in the light of the remarkable recent discoveries of observational cosmology and the enthusiastic advocacy of theoretical cosmologists and the particle physics community.

5.2 The thermal history of the Universe

First of all, let us develop a much more complete thermal history of the Universe than that discussed in Section 4.8. In that section, we established the important result that the temperature of the cosmic microwave background radiation changes inversely as the scale factor R, $T_{rad} \propto R^{-1}$, and so, as we take the Universe backwards in time, the radiation temperature increases. We recall that the scale factor R describes how the typical distance between galaxies changes as the Universe expands and it is related to the redshift z by $R = 1/(1 + z)$. As we showed in Section 4.8, since the radiation temperature of the cosmic microwave background is 2.725 K at the present epoch, which corresponds to $R = 1$ and $z = 0$, the scale factor had the value $R = \frac{1}{2}$ at a redshift $z = 1$ and the temperature of the background radiation was about 5.45 K. In Section 4.8, we extrapolated this relation back to the time when the scale factor was only about one thousandth of its present value and the temperature of the background radiation was high enough to ionise all the neutral hydrogen present in the Universe. Prior to this epoch, the *epoch of recombination*, the temperature of the radiation continued to increase as $1/R$ and the primordial gas was fully ionised.

We now need to study the physics of the *pre-recombination era*, because, although we cannot observe them directly, some cosmic fossils have survived from very much earlier times. As we squash the Universe backwards in time, the density of matter increases as the inverse cube of the scale factor, $1/R^3$, exactly as expected if a sphere of matter is compressed. At the same time, the temperature of the cosmic background radiation increases as $T \propto 1/R$. One of the most important discoveries of 19th century physics was the fact there is a simple relation between the amount of energy contained in a black-body radiation spectrum and its radiation temperature. This famous law, known as the *Stefan–Boltzmann law*, states that the energy contained in black-body radiation is proportional to the fourth power of the temperature. Now, the spectrum of the cosmic background radiation has a perfect black body form and so we know precisely how the energy, or more precisely the energy density, of the radiation changes with cosmic epoch. Since the temperature of the cosmic background radiation increases with scale factor, $T \propto 1/R$, it follows that the energy density, u, of the radiation increases with decreasing scale factor R as the inverse fourth power of the scale factor, $u \propto 1/R^4$. This is a key result for cosmology because it means that, as the Universe is squashed further and further into the past, the energy density in the radiation increases more rapidly than the density of matter. The reason this is so important is that Einstein's theory of special relativity tells us that the energy present in radiation has a certain mass, according to Einstein's famous relation $E = mc^2$. Thus, the mass density associated with the radiation increases more rapidly with decreasing R than does the mass density in the ordinary matter. If we go far enough back in time, the mass density in the radiation eventually exceeds that in the ordinary matter and then the dynamics of the Universe are determined by the mass, or energy density, of the radiation rather than by that of the matter. The dynamics of the late Universe are dominated by the mass density of the matter and these phases are known as the *matter-dominated* phases; the dynamics of the early Universe are dominated by the mass density of radiation and these phases are known as the *radiation-dominated* phases.

The epoch at which the dynamics changed from being radiation- to matter-dominated depends upon the value of the density parameter but, adopting the critical density $\Omega = 1$ for the sake of definiteness, the transition took place when the scale factor R was about 1/10 000. Thus, for the case of the critical Universe, the epoch at which this transition took place occurred well before the epoch of recombination. The dynamics of the radiation-dominated phases are quite different from those of the matter-dominated phases, for which the variation of the scale factor R with cosmic time was illustrated in Figure 4.9. During the radiation-dominated epochs, the scale factor R changes as the square root of the cosmic time t, $R \propto t^{\frac{1}{2}}$. The matter is in the form of a fully ionised plasma and is tightly coupled to the radiation by scattering, so that, as the Universe expands and cools, the radiation and the matter cool at the same rate.

As we go yet further back in time, the next critical stage occurs when the Universe was squashed to about 1/300 000 000 of its present size. You may ask, 'Doesn't the Universe become quite fantastically dense when it is squashed by this enormous factor?' Perhaps surprisingly, the answer is 'No!'. The reason is that the average density of matter in the Universe at the present time is really very, very low indeed. When the Universe was squashed by a factor of 300 000 000, the average density of the matter in the Universe was still much less than typical room densities. The behaviour of matter at such densities is well understood, the only difference being that the matter and radiation are now at a temperature of about 1 000 000 000 K. Although this is a high temperature, it is still modest on the scale of the energies studied in particle physics experiments. The elementary processes which occur in such high temperature plasmas are well understood, as are the nuclear interactions which are expected to take place. This thermal history of the Universe is summarised in Figure 5.1.

This temperature of 1 000 000 000 K is a very interesting temperature – it is so high that the cosmic background radiation spectrum is shifted to γ-ray energies and these γ-rays are sufficiently energetic to dissociate the nuclei of atoms into their constituent protons and neutrons. In other words, the nuclei of any chemical elements which happened to be present at that time could not survive, because they would be broken up into protons and neutrons by the thermal background radiation. At even earlier times, the temperature continues to increase with decreasing scale factor and collisions between the particles can create the whole zoo of elementary particles when the temperature becomes high enough. We will have to take these very early phases seriously in a moment but let us first explain why cosmologists are so confident that the Universe actually passed through this hot dense phase.

5.3 The synthesis of the light elements

In 1964, while I was completing the first year of my research as a graduate student at Cambridge, Fred Hoyle gave a memorable course of lectures on the problems of extragalactic research. He would arrive with a scrap of paper with a few notes and expound an area of current research. One week, the topic was the problem of the cosmic helium abundance. Helium is one of the more difficult elements to observe astronomically because it can only be observed in very hot stars. By 1961, it was becoming clear that the abundance of helium

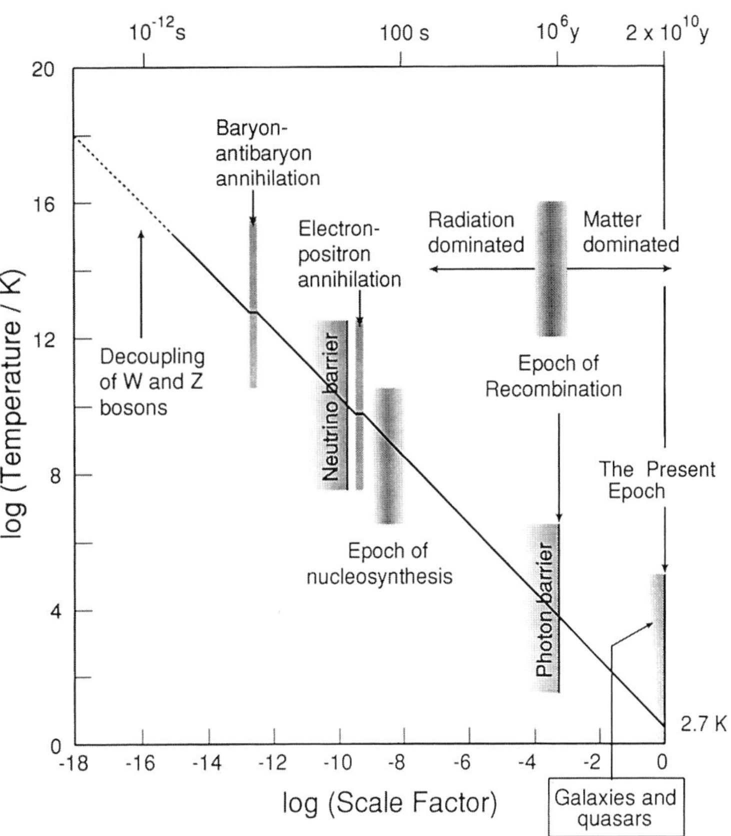

Figure 5.1. The thermal history of the Universe. The radiation temperature decreases as $T_r \propto 1/R$, except for abrupt jumps as different particle–antiparticle pairs annihilate. Various important epochs in the standard Big Bang model are indicated. An approximate time-scale is shown along the top of the diagram. The photon and neutrino barriers are shown – in the standard Big Bang picture, the Universe is opaque for electromagnetic radiation and neutrinos prior to these epochs.

seemed to be remarkably uniform wherever it could be observed, its percentage abundance by mass corresponding to about 25%. A further important observation, reported by Robert O'Dell in 1963, concerned the helium abundance in a planetary nebula in the old globular cluster M15. Despite the fact that the heavy elements were significantly depleted relative to typical cosmic abundances, the helium abundance was still about 25% by mass. Hoyle reviewed the evidence on the cosmic helium abundance and then described the work of George Gamow and his colleagues Ralph Alpher, Robert Herman and James Follin, who had studied the problem of the synthesis of the elements in the hot early phases of the Big Bang in the late 1940s and the early 1950s. They worked out the thermal history of the radiation-dominated Universe, which was described in the last section, but they did not succeed in explaining the synthesis of the chemical elements by primordial nucleosynthesis. In 1953, Hoyle discovered the triple-α process, which is the key interaction for synthesising carbon from three helium nuclei. Over the next few years, Hoyle and his colleagues Geoffrey and Margaret Burbidge and William Fowler, showed how the elements could be synthesised in the interiors of stars. Interest in primordial nucleosynthesis evaporated until the large abundance of helium was established.

By 1964, it was possible to carry out calculations of the synthesis of the elements in the early phases of the Big Bang more accurately using digital computers. At that time, Roger Tayler had just returned to Cambridge and was present in the audience. Hoyle and Tayler realised that they could undertake more precise calculations than had been undertaken previously and, over the next two weeks, they and Tayler's research student John Faulkner worked out the details of nucleosynthesis in the early phases of the Big Bang. The audience had the privilege of being present as a key piece of modern cosmology was created in real time in an undergraduate lecture course. Hoyle and Tayler obtained the answer that about 25% helium by mass is synthesised in the Big Bang, in remarkable agreement with observation and essentially independent of the present matter density in the Universe. Their paper on this subject was published in *Nature* in 1964.

This is such an important topic that it is worth looking into the results of these calculations in a little more detail. Following these pioneering calculations, Hoyle went on to undertake a much more complete analysis of primordial nucleosynthesis in collaboration with Robert Wagoner and William Fowler. What they did was to evolve model Universes from states of very high density and temperature, including all the best nuclear physics available concerning the interactions between all types of particles and nuclei. Figure 5.2 illustrates the results of some of these classic computations and shows how the abundances of various particles and nuclei change through the critical period when the Universe was only a few minutes old.

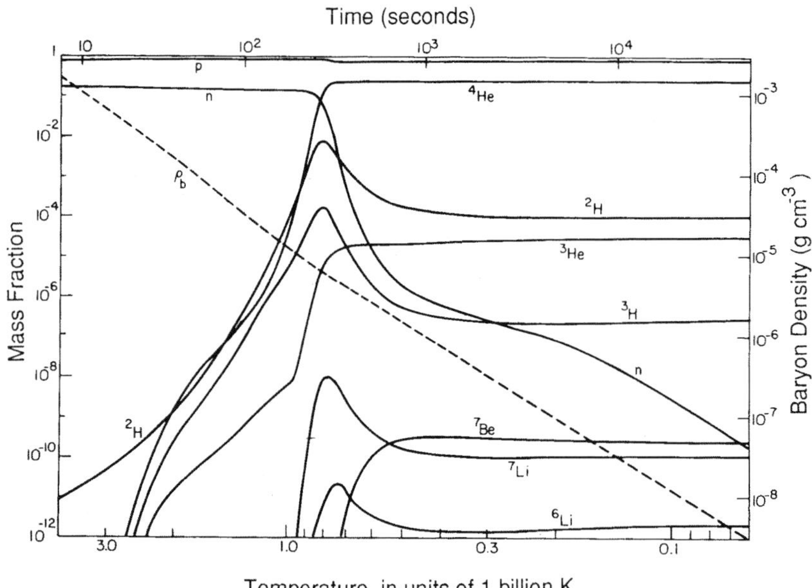

Figure 5.2. A diagram showing how the abundances of the chemical elements are expected to change during the first 30 minutes of the Big Bang. Time measured from the Big Bang is shown along the top of the diagram and the corresponding temperature along the bottom. The diagram shows how the abundances by mass (or mass fractions) of the different elements build up as the Universe cools down.

Let us study the evolution of the various constituents of the Universe, as illustrated in Figure 5.2. The computations begin at such high temperatures and densities that all the particles which can exist at these temperatures are maintained in their equilibrium abundances. Protons, neutrons and radiation are present, as well as other particles which must be present because the system is in thermal equilibrium – these particles include neutrinos, muons, electrons and their antiparticles. Of particular importance is the fact that the protons and neutrons are maintained in their equilibrium abundances by weak interactions involving electron neutrinos

$$e^+ + n \leftrightarrow p + \bar{\nu}_e \qquad \nu_e + n \leftrightarrow p + e^-$$

As the Universe cools, the time-scales for these weak interactions increase rapidly and, when the Universe is about one second old, they become greater than the age of the Universe. The neutrinos are therefore no longer able to maintain the equilibrium abundances of neutrons and protons at later times and the neutron-to-proton ratio is more or less frozen at the value it had when the neutrinos decoupled. This epoch, when the neutrinos decoupled from the protons and neutrons, results in a *neutrino barrier*, similar to the last scattering surface for electromagnetic radiation, which occurs at the epoch of recombination. If we were able to observe the Universe in neutrinos, we would never be able to observe neutrinos from epochs earlier than this neutrino barrier, which is shown in Figure 5.1.

The electrons and their antiparticles, the positrons, are also maintained in their equilibrium abundances by the process of electron–positron pair production in collisions of high energy γ-rays of the thermal background radiation but, when the average energy of the γ-rays falls below the rest mass energy of the electron, the electrons and positrons annihilate and the energy released slightly heats up the thermal background radiation, as indicated by the little jump in the temperature history of the Universe shown in Figure 5.1.

All that are left now are protons and neutrons, which can combine to form the simplest compound nucleus, deuterium (^2H), which consists of a proton and a neutron – this simple nucleus is often referred to as a deuteron. When the radiation temperature is as high as 3 000 000 000 K, however, the deuterons do not survive as they are dissociated by the γ-rays of the background radiation and a very low abundance of deuterium is expected, as can be seen from Figure 5.2. As the Universe continues to expand and cool down, however, more and more of the deuterium survives. The deuterons can then interact with protons, neutrons and other deuterons to form heavier elements. By the time the temperature has dropped to 300 000 000 K, all the nuclear reactions are over. Inspection of Figure 5.2 shows that most of the neutrons have been combined with protons to form the most stable helium isotope, ^4He, but there are also traces of other light elements such as deuterium, the lighter isotope of helium, ^3He, and a little lithium, ^7Li. It is striking that no heavier elements such as carbon and oxygen are created. The reason for this is that there are no stable species with five and eight nucleons – beryllium-8, for example, is a totally unstable nucleus.

The spectacular result of these calculations is that the elements which are formed by this process of *primordial nucleosynthesis* are precisely those which have proved impossible to account for by nucleosynthesis in stars. As Hoyle and Tayler pointed out, it is a problem to understand why, wherever we look in the Universe, helium always seems to be present with an abundance of about 25% by mass, quite independent of the abundance of the heavier elements. The remarkable result of these calculations of primordial nucleosynthesis is that, for any reasonable value of average density of ordinary matter in the Universe now, about 23 to 24% of helium by mass is created by nuclear reactions in the first few minutes of the Big Bang. While stars can produce a small amount of helium, we do not know of any way of creating as much as 25% of helium by nucleosynthesis in stars. Equally, it is very difficult to understand how deuterium could be created by nucleosynthesis inside stars. It is such a fragile element that it is destroyed in stellar interiors rather than being created. Primordial nucleosynthesis in the Big Bang solves these dilemmas.

The other great realisation has been that these light elements are important diagnostic tools for the average density of ordinary matter in the Universe. Deuterium, the heavy isotope of

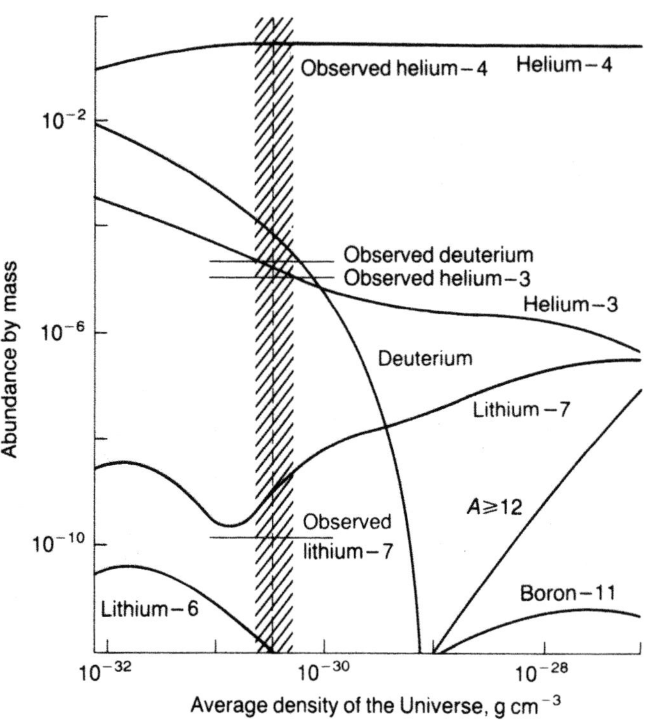

Figure 5.3. The observed abundances of the light elements compared with the predictions of different standard world models. The solid lines show the predicted abundances for different assumed values of the density of ordinary matter. A range of low density models can account for the observed abundances of helium-4, deuterium, helium-3 and lithium.

hydrogen, is of special importance, because it cannot be created by nucleosynthesis in stars. It is such a fragile nucleus that, if it is present in the nuclear burning regions in the centres of stars, it is immediately destroyed by high energy γ-rays. In other words, we never expect stars to make any deuterium at all, rather they can only destroy it. In the Big Bang, significant amounts of deuterium are produced because of the rapid expansion of the Universe in its early phases – there is not enough time to destroy all the deuterium by γ-rays or to convert it all into helium by further nuclear reactions. It turns out that the deuterium abundance is a very sensitive probe of the density of ordinary matter in the Universe today. If the density of ordinary matter were high, there would be sufficient collisions between deuterium and other nuclei to convert almost all of it into helium. If, however, the density of ordinary matter is low, there is not enough time for all the deuterium to be converted into helium, leaving some deuterium over, which becomes the cosmic abundance of deuterium we observe today. The dependence of the amount of deuterium produced upon the average density of ordinary matter in the Universe is displayed in Figure 5.3.

The argument then proceeds as follows – the more deuterium created in the Big Bang, the lower the average density of ordinary matter in the Universe must be. Since we only know of ways of destroying deuterium, the present abundance of deuterium, which amounts to about one part in 100 000 by mass of the abundance of hydrogen, sets a rather firm upper limit to the density of ordinary matter in the Universe. If the density were any greater, not enough deuterium would be produced. The importance of this result is that it tells us that the density of ordinary matter in the Universe must be less than about one tenth of the critical density. In fact, it turns out that we can explain the observed abundances of helium-3 and lithium-7 as well for a single value of the present mass density of ordinary matter in the Universe (Figure 5.3). Recent analyses have suggested that the ordinary matter density probably amounts to only about 2–6% of the critical density.

What these calculations tell us is that, if our Universe really has the critical density, and we will argue that there are good reasons to suppose it probably has, then most of the matter in the Universe cannot be in the form of ordinary matter. If $\Omega = 1$, then most of the matter must be present in some dark, unknown form which is not made up of ordinary matter. This is another incarnation of the dark matter problem which was discussed in Section 4.5. The dark matter may well be in some exotic form, such as the types of weakly interacting massive particle predicted by theories of elementary particles. This is why studies of the very early Universe have excited so much interest among the particle physicists who regard it as a laboratory for testing theories of elementary particles.

The importance of these studies is that they provide us with convincing evidence that the Universe indeed passed through a very hot, dense phase. The helium-4, deuterium, helium-3 and lithium are fossils laid down in the first ten minutes of the Big Bang and we find them all around us at the present epoch.

An intriguing footnote to this story concerns the fact that the predicted abundances of the light elements are sensitive to the dynamics of the Universe at the time when the processes of nucleosynthesis took place. In particular, if the expansion were more rapid than the standard model, the freezing of the proton–neutron ratio would occur at earlier times and result in the overproduction of helium. This argument enables useful constraints to be placed upon a number of cosmological parameters. For example, we can exclude the possibility that the gravitational constant was greater in the past because then the Universe would have had to expand much more rapidly at early times to attain its present size. Equally intriguing is the fact that limits can be set to the number of different species of neutrino which can be present in the Universe. Three different species of neutrino are known, the electron, muon and tau neutrinos, and if there were any more of them, they would contribute to the total energy density of the Universe at that time and so speed up the expansion through the period of primordial nucleosynthesis. The limits to the cosmic abundance of helium enabled the astronomers to show that there can only be three types of neutrino. This result was subsequently confirmed by particle physics experiments at the Large Electron–Positron (LEP) collider at CERN.

5.4 Nine facts about the Universe

Let us now try to pull together all the information we have been assembling about the nature of our Universe over the last $4\frac{1}{2}$ chapters. From the mass of observations we have discussed, which of them provide information of real cosmological significance?

When I began research in radio astronomy as a graduate student in 1963, my supervisor Peter Scheuer handed me a copy of Hermann Bondi's classic text *Cosmology* to absorb and warned me that:

> There are only $2\frac{1}{2}$ facts in cosmology.

The point was an important one in that, of the mass of observations which can be made of gas, stars and galaxies, most of them tell us nothing of real cosmological significance. We have to select from the plethora of data those pieces which establish real facts about the nature of our Universe as a whole.

In 1963, the $2\frac{1}{2}$ facts were as follows:

Fact 1. *The sky is dark at night.*

This is a familiar but profound observation which leads to what is known as *Olbers' paradox*, although the paradox was well known to earlier cosmologists. In its simplest form, the paradox arises because, if the Universe were infinite, static and uniformly filled with stars, the sky would be as bright as the surface of the stars, clearly in contradiction with our experience. At least one of these assumptions about the nature of our Universe has to be wrong. Bondi gives a thought-provoking discussion of the significance of the paradox in his book *Cosmology*. The fact that the sky is not as bright as the surface of the Sun provides us with some very general information about the Universe. Probably the most general way of expressing the significance of this observation is that the Universe must, in some sense, be far from equilibrium, although the way in which it is in disequilibrium cannot be deduced from this very simple observation. The fact that the Universe is expanding and has a finite age are two contributions to the resolution of the paradox.

The second fact was Hubble's law, which we discussed in some detail in Section 4.2.

Fact 2. *The galaxies are receding from our own Galaxy and their velocities of recession are proportional to their distances from our Galaxy,* $v = H_0 r$.

The half fact concerned the number counts of extragalactic radio sources which had shown that there are many more faint radio sources than would be expected according to the standard Friedman world models and to steady state cosmology, which had attracted considerable attention at that time. Martin Ryle had interpreted these data as showing that there must have been more extragalactic radio sources in the distant past than there are now. In other words,

Fact $2\frac{1}{2}$. *The contents of the Universe have changed as the Universe grows older.*

The reason for the ambiguous status of this fact was that, at that time, the number counts of extragalactic radio sources were a matter of considerable controversy, particularly with the proponents of steady state cosmology. I was plunged into the middle of that debate as soon as I began my research programme with Martin Ryle and Peter Scheuer. As we have discussed above, this is no longer a controversial issue – many classes of object are now known to exhibit evolutionary changes as the Universe grows older.

Thus, in 1963, the number of real facts which characterised the Universe as a whole was very small and modest progress had been made since the 1930s. The three decades since then have been a golden age for astrophysics and cosmology as one discovery has succeeded another in rapid succession.

Let us now add to this list the new facts which have come to light since 1963. Facts 3 and 4 resulted from the discovery of the cosmic microwave background radiation by Arno Penzias and Robert Wilson in 1965. The properties of the cosmic microwave background radiation were described in some detail in Sections 1.6 and 4.2 and they unquestionably provide information about the properties of the Universe as a whole. Specifically, the results of the COBE mission show that:

Fact 3. *The Universe is isotropic on very large scales to an accuracy of better than one part in 100 000.*

As a footnote to Fact 3, we can note that tiny fluctuations have now been detected in the spatial distribution of the cosmic microwave background radiation at a level of about one part in 100 000. These are of central importance for the study of galaxy formation, as we discussed in Chapter 4.

Fact 4. *The spectrum of the cosmic microwave background radiation is of pure black-body form at a radiation temperature of 2.725 K.*

We discussed experimental and observational tests of the general theory of relativity in Section 3.6 and also evidence that the gravitational constant has not changed over cosmological time-scales in that section and in Section 5.3. These considerations indicate that general relativity is by far the best theory of gravity we possess and is the correct starting point for the construction of cosmological models.

Fact 5. *Standard general relativity has passed the most precise tests which have been devised so far and there is no evidence that the gravitational constant has changed over cosmological time-scales.*

The half fact which Peter Scheuer warned me about in 1963 is no longer in question. All types of extragalactic radio source, all classes of quasar and the faint X-ray sources surveyed by the ROSAT X-ray Observatory have shown clear evidence for large changes in these populations with cosmic epoch. Deep counts of galaxies carried out in the optical waveband show that there is an excess of faint blue galaxies and this is interpreted in general terms as a change in the numbers of these galaxies as the Universe grows older. We can therefore conclude that:

Fact 6. *Many different classes of extragalactic system show changes in their average properties with cosmic epoch.*

The dark matter problem is so pervasive in extragalactic astronomy that it can be safely elevated to the status of a fact.

Fact 7. *Most of the mass of the Universe is in some dark form and its mass exceeds that in the form of visible matter by at least a factor of 10.*

It is important to note that there are two different routes which lead to the dark matter problem. The first of these we can call *astrophysical* dark matter, in the sense that we find that there must be dark matter present in galaxies and clusters of galaxies in order to account for their dynamical stability. Dynamical evidence also suggests that dark matter is present on even larger scales. This type of evidence should be contrasted with what we can call *cosmological* dark matter, in other words, dark matter which must be present if the Universe is to have density parameter $\Omega = 1$. We have already encountered one line of reasoning in favour of this value of Ω, namely, the difficulty of understanding how galaxies and other large scale structures could have formed by the present day without producing excessively large fluctuations in the cosmic microwave background radiation (Sections 4.7 and 4.8). We will discuss other lines of argument in favour of the critical model in Section 5.5.

The agreement between the observed light element abundances and the predictions of the standard Big Bang picture is very impressive and is worthy of being elevated to Fact 8.

Fact 8. *The light elements, helium-4, deuterium, helium-3 and probably lithium, were created by primordial nucleosynthesis in the early phases of the Big Bang. A consequence of this result is that the density of ordinary matter in the Universe must be at least a factor of 10 less than the critical density.*

Finally, the large scale structure of the Universe has now been defined in sufficient detail that we can raise its 'sponge-like' structure on the large scale to the status of a fact.

Fact 9. *The distribution of galaxies on large scales in the Universe, although uniform on the cosmological scale, possesses large scale irregularities on scales much greater than those of clusters of galaxies.*

The acceptability of any physical theory depends upon the economy with which it can account for independent pieces of observational or experimental evidence. The case for the Big Bang model for our Universe rests upon the fact that it can account naturally for a number of the above independent pieces of evidence. The standard theory is based upon the assumption that the large scale dynamics of the Universe can be described by the general theory of relativity and upon the application of physics which has been tried and tested in the laboratory to the Universe at large. We need the important clue from observation that we can consider the Universe to be isotropic and homogeneous on a large scale. Then, the standard Big Bang model can naturally account for the following features of our Universe:

(1) The Hubble expansion of the Universe.
(2) The thermal spectrum and isotropy of the cosmic microwave background radiation.
(3) The abundances of the light elements.

These are independent observations which can be reconciled with a single Big Bang model of the Universe. Facts 6, 7 and 9 are not in conflict with the standard isotropic Big Bang model. We need to understand much more about the astrophysics of the formation and evolution of galaxies to address Fact 6 in a serious way. While Facts 7 and 9 can be accommodated within the standard Big Bang picture, they lead to fundamental problems. In fact, there are four of these and they are the subject of the next Section. Despite these shortcomings, the evidence presented above has persuaded virtually all cosmologists that the Big Bang picture is the most convincing theoretical framework within which to carry out cosmological research.

5.5 The four fundamental problems

It is remarkable that so much progress has been made in understanding the past history of our Universe. The success of the standard Big Bang in accounting for the cosmic abundances of the light elements by primordial nucleosynthesis suggests that we can trust the model as far back in time as a fraction of a second from the beginning of the Big Bang. Most cosmologists find the picture so convincing that it has become the standard framework for cosmological research. Like all good theories, however, it raises as many problems as it solves and the solutions to these must lie in the very early Universe. There are four fundamental problems with the standard picture and we have already dealt with one of these in some detail in Chapter 4, namely, the origin of the density fluctuations from which galaxies and the large scale structure of the Universe formed. According to the analysis of Chapter 4, these fluctuations have to be included in the initial conditions from which the Universe evolved if galaxies are to be formed by the present day – this is scarcely a satisfactory physical picture. The three other basic problems are of the same nature.

5.5.1 The horizon problem The remarkable isotropy of the Universe on a large scale poses a knotty problem for the standard Big Bang picture. The COBE observations of the cosmic microwave background radiation have shown that regions in opposite directions on the sky have exactly the same intensity and spectrum to a precision of one part in 100 000. The big problem is 'How was it that these regions came to have exactly the same properties, despite the fact that they cannot communicate with each other?' According to conventional thinking, if one region of the Universe is to have the same properties as another region, at the very least, the two regions must be able to communicate with each other, since otherwise they cannot 'know' that they have to be the same and adjust their properties accordingly. Now, the fastest that information can travel is the speed of light. The problem arises because there is only a finite amount of time since the origin of the Big Bang.

As we go further and further back in time, the distance over which information can be communicated becomes smaller and smaller. As an example, the microwave background radiation originates from a last scattering surface at the epoch of recombination when the Universe was contracted by a factor of about 1000 as compared to its present size. We can work out how far a light ray could have travelled along the last scattering surface since the beginning of the Universe and convert that distance into an angle on the sky. This angle turns out to be only 5° (Figure 5.4). This means that there is no way in which regions in opposite directions on the sky could have been in causal communication. How then could the different regions of the Universe know that they had to end up looking the same in all directions? This problem is known as the *horizon problem*. The only way of accounting for the isotropy of the Universe according to the standard picture is to assume that the Universe was set up in that way in the first place. In other words, we have to assume that the initial conditions from which the Universe evolved were isotropic. This seems a rather arbitrary assumption.

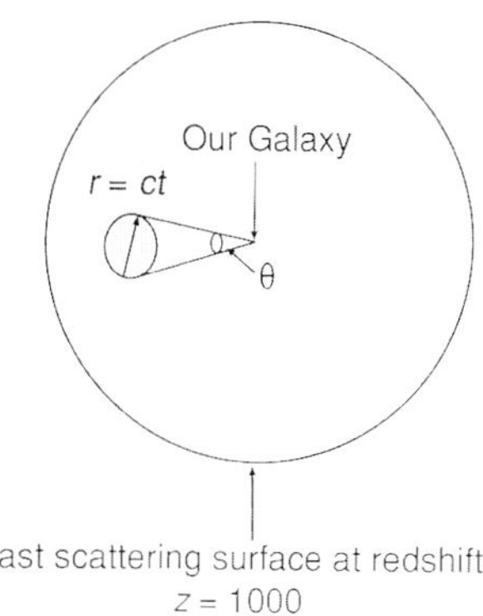

Figure 5.4. Illustrating how far light can travel along the last scattering surface at a redshift of about 1000 in the time since the origin of the Big Bang. The distance light travels corresponds to an angle of only 5° on the sky and so most of the sky cannot have been in causal contact. The regions which can be in causal contact become smaller and smaller at larger redshifts.

5.5.2 The flatness or fine-tuning problem The next problem arises from the fact that our Universe is within a factor of ten of the critical density, $\Omega = 1$. In Section 4.5, we described how, when we take account of the dark matter in galaxies and clusters of galaxies, the density parameter is probably within about 10 to 20% of the critical density. There may well be even more dark matter on larger scales and so our Universe must be quite close to the critical model. Now, according to the standard Big Bang models, there is no reason why the density parameter should take any particular value at all. If we were to select Universes at random, there is no reason why the density parameter should be close to the critical value. In principle, our Universe might have taken values of the density parameter which could be have been billions of times greater or less than the critical value. There is nothing in the physics of the standard models which tells us what the density parameter should be.

The problem is, however, much more severe than this because we can show that, if the value of the density parameter departed even very slightly from the critical value in the very early Universe, then it would depart from the critical value at the present day by an enormous factor. To express this problem in another way, all Universes with values of Ω which are not equal to one are unstable, in the sense that they diverge more and more from the case $\Omega = 1$ as time goes on.

The simplest way of illustrating this problem is to study the behaviour of rockets as they escape from the Earth. Suppose the rocket starts off with a velocity which is just greater than the escape velocity (see Section 4.3 and Figure 4.10). As the rocket travels further and further away from the Earth, the escape velocity at that point becomes less than it was on the surface of the Earth, because the rocket is further away from the Earth. However, the ratio of the velocity of the rocket to the local escape velocity becomes greater and greater as the rocket moves further away from the Earth. In Figure 5.5, the ratio of the velocity of the rocket to the escape velocity is plotted as the rocket travels further and further from the Earth. It can be seen that the divergence from the escape velocity becomes enormous. Furthermore, even if the initial velocity was only slightly greater than the escape velocity, the divergence takes place if we wait long enough. It follows that, if we observe a rocket at a very great

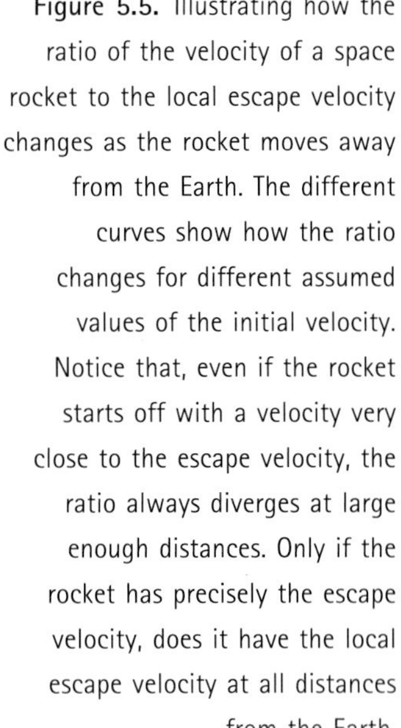

Figure 5.5. Illustrating how the ratio of the velocity of a space rocket to the local escape velocity changes as the rocket moves away from the Earth. The different curves show how the ratio changes for different assumed values of the initial velocity. Notice that, even if the rocket starts off with a velocity very close to the escape velocity, the ratio always diverges at large enough distances. Only if the rocket has precisely the escape velocity, does it have the local escape velocity at all distances from the Earth.

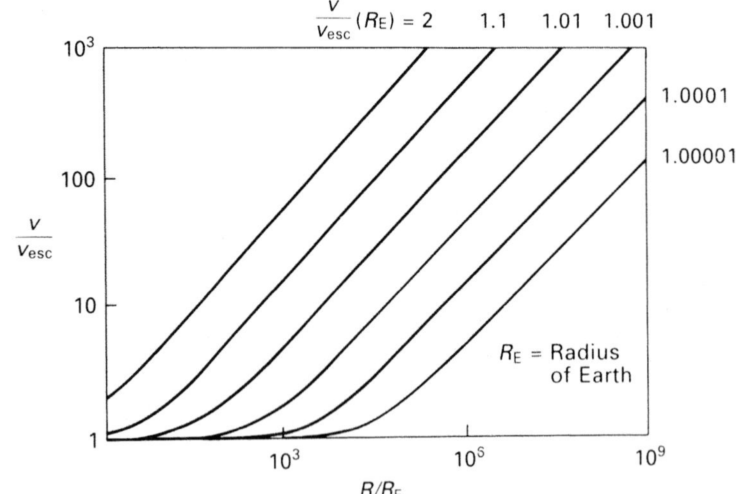

distance from the Earth and it has more or less the escape velocity at that point, it must have started out with almost exactly the escape velocity with quite remarkable precision.

The Universe has exactly the same problem – unless it was set up with more or less exactly its own escape velocity at the beginning, it cannot end up with an expansion velocity close to its escape velocity now. The significance of this result is apparent – the Universe must have been set up in the beginning with almost exactly the critical density with quite remarkable precision, if it is to end up being within a factor of ten of the critical density now. This is sometimes called the *fine-tuning problem*. It is also known as the *flatness problem*, because the critical model with $\Omega = 1$ has flat spatial geometry. Despite the fact that, in the standard Big Bang, there is no reason why our Universe should have been set up with the critical density, it must have been set up in that way in its very early stages. Theorists argue very strongly that, because of the fine-tuning problem, the density parameter of the Universe must take the value $\Omega = 1$. As we have argued, there is no observational reason to reject this value, provided there is sufficient dark matter in the space between rich clusters of galaxies.

5.5.3 The baryon asymmetry problem The last problem results from considerations of what happens when we extrapolate to higher and higher temperatures in the early Universe. One of the great discoveries of the 1930s was that, for every type of particle, there exists a corresponding antiparticle, which is its mirror image. For example, we know that, in collisions between high energy γ-rays and ordinary matter, large numbers of electrons and their mirror-image particles, the positive electrons or positrons, are created. Figure 5.6 is a sketch of the tracks observed in a particle physics experiment in which an electron and positron have been created. The electrons are negatively charged particles and the positrons are identical but with positive charges and so their paths are bent in opposite directions by the magnetic field. All the properties of antiparticles are reversed as compared with those of particles. In exactly the

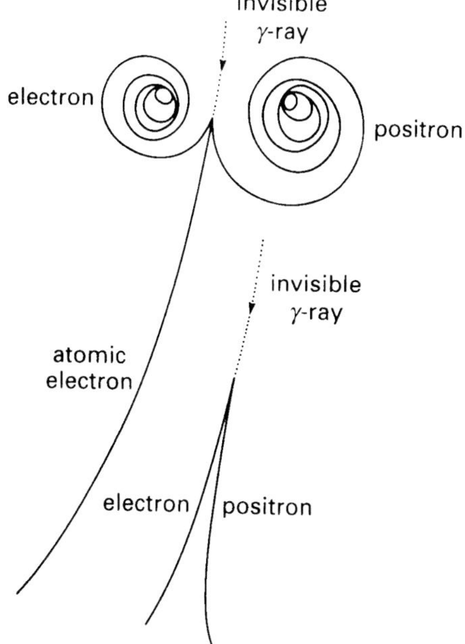

Figure 5.6. An example of an electron–positron pair production event taking place in a particle physics experiment. The electrons and positrons have opposite charges and so are bent in opposite directions by the strong magnetic field in the detector.

same way, protons have a positive electric charge and their mirror image particles, the antiprotons, have negative charge.

If we extrapolate the Big Bang model back to the epochs when the temperature was about 10^{12} K, the thermal background radiation consists of very high energy γ-rays. The high energy γ-rays collide, producing equal numbers of protons and antiprotons, provided enough energy is present to create the particle–antiparticle pair. This is another application of Einstein's mass–energy relation, $E = mc^2$ – in other words, the high energy γ-rays have to have enough energy to create the masses of both the particle and the antiparticle. This results in a rather curious situation in the early Universe because we know that locally our Universe is made exclusively out of matter, rather than out of an equal mixture of matter and antimatter. There must be very little antimatter present in the Universe today or it would have been recognised by the observation of the γ-rays which would be emitted when it annihilates with matter.

It is straightforward to work out how many γ-rays there were in the early Universe relative to the number of particles of matter – it turns out that there were about 100 million to 1 billion γ-rays for every proton. Therefore, when the temperature became sufficiently high for the creation of protons and antiprotons from collisions of high energy γ-rays, the Universe was suddenly flooded with millions of proton–antiproton pairs. At these very early times, there must have been, say 1 000 000 001 protons for every 1 000 000 000 antiprotons. As the Universe cooled down, the 1 000 000 000 protons annihilated with the 1 000 000 000 antiprotons producing pairs of γ-rays, leaving only one proton which has the correct γ-ray to proton number ratio. This ratio was preserved as the Universe expanded and became the ratio of the number of photons of the cosmic microwave background radiation to the number of protons at the present day. If the Universe was initially completely symmetric between matter and antimatter, almost all the matter would have been annihilated with antimatter. Detailed calculations have shown that we would have ended up with a very small amount of matter and antimatter in the Universe today, far less than the amount of matter we observe by a factor of about 1 000 000 000. Hence, in the standard Big Bang picture, a slight asymmetry between the numbers of protons and antiprotons, or more generally, between the particles and antiparticles, has to be present in its very early stages. Within the framework of the standard Big Bang, this initial condition has to be introduced arbitrarily in order to 'explain' the observed photon-to-proton ratio today.

5.6 Solutions to the great problems

We have now completed our catalogue of the four great problems of Big Bang cosmology. To recapitulate, these are:

(1) Why is the Universe isotropic?

(2) Why is the Universe so close to its critical density?

(3) Why was the Universe slightly asymmetric with respect to matter and antimatter in its early phases?

(4) What is the origin of the density fluctuations out of which galaxies formed?

In the standard Big Bang model, these four problems are solved by assuming that the Universe is endowed with the correct initial conditions to begin with. In other words, we have to postulate that the initial conditions for our Universe were such that it was isotropic, that it was set up with density very, very close indeed to the critical density, that there was a slight matter–antimatter asymmetry, and that there were fluctuations present out of which galaxies and the large scale structure of the Universe were eventually to form. Scientifically, this is not a very satisfactory picture because it has no explanatory value – we only get out at the end what we put in at the beginning.

Let us consider five approaches to solving these problems.

(1) That is just how the Universe is – the initial conditions were set up in that way.

(2) There are only certain classes of Universe in which intelligent life can evolve. The Universe has to have the appropriate initial conditions and the fundamental constants of nature should not be too different from the values they have today or else there would be no chance of life forming as we know it. This approach is known as the *anthropic cosmological principle* and it asserts that the Universe is as it is because we are here to observe it.

(3) The inflationary scenario for the early Universe.

(4) Seek clues from particle physics and extrapolate that understanding beyond what has been confirmed by experiment to the earliest phases of the Universe.

(5) Something else which we have not yet thought of. This would certainly involve some new physical concepts.

There is some merit in each of these approaches. For example, even in the case of the somewhat pessimistic approach (1), it might turn out to be just too hard a problem to disentangle the physics responsible for setting up the initial conditions from which our Universe evolved. How can we possibly check that the physics adopted to account for the properties of the very early Universe is correct?

5.7 The anthropic cosmological principle

As McCrea pointed out in 1970, part of our problem stems from the fact that we have only one Universe to study and that is the one we live in. We cannot go out and investigate other Universes to see if they have evolved in the same way as ours. Therefore, we cannot exclude the possibility that our Universe has these properties just by chance. There is an intriguing line of reasoning which states that there are only certain types of Universe in which life as we know it could possibly have formed. For example, the stars must live long enough for biological life to form and evolve into sentient beings which can even ask the question. This line of thinking is formalised in what is known as the *anthropic cosmological principle*. In the construction of cosmological models, it is assumed that the Earth and our Galaxy are not located at any privileged position in the Universe. This is a natural extension of the Copernican principle to the Universe as a whole. The idea that we are not located at any special position in the Universe is known as the *cosmological principle* and has been implicitly assumed in the construction of the standard world models.

Now, it is clear that we are to a considerable degree privileged by being able to undertake the scientific study of astrophysics and cosmology at all. The intriguing question is to what extent this special position is relevant in scientific studies of the Universe. The problem is dealt with in some detail in the excellent book *The Anthropic Cosmological Principle* by John Barrow and Frank Tipler. Another version of the same type of consideration is contained in the book *Cosmic Coincidences* by John Gribbin and Martin Rees.

Maybe there is some truth in these ideas and, out of all possible Universes which could have formed, we live in the one which has exactly the correct initial conditions and the correct values of the physical constants of nature so that human life can develop to such a state that we pose the fundamental questions and I can write this book. I confess that I do not like this line of reasoning, because the logical consequence is that we will never be able to find any physical reason for the initial conditions from which the Universe evolved or for the relations between the fundamental constants of nature. I regard the anthropic cosmological principle as the very last resort if all other physical approaches fail. I believe we are very far from that position yet.

5.8 The inflationary Universe and clues from particle physics

Fortunately, we now have clues which suggest ways in which these problems may be solved. Much of the stimulus for these new ideas has come from discoveries in particle physics. To do justice to these would take us far into the complexities of modern particle physics and so let us be selective and introduce some concepts which give me hope that we may eventually be able to solve the fundamental problems in a unified picture for the physics of the early Universe. The big problem is that the energies at which many of the most important processes are believed to take place far exceed those attainable in the most powerful particle accelerators. Consequently, much of the analysis is speculative and based upon extrapolations of the theory of elementary particles far beyond the energies at which it has been fully validated. Nonetheless, the procedure of using astronomical and cosmological arguments to gain a deeper understanding of physical principles has a long and distinguished pedigree. Perhaps the greatest example was Newton's derivation of his law of gravity from Kepler's laws of planetary motion, which, in turn, were derived from Tycho Brahe's magnificent observations of the orbits of the planets. In the same way, the late 20th century cosmologist uses the early Universe as a laboratory within which to test and constrain theories of elementary particles at energies unattainable in terrestrial laboratories.

Let us discuss first of all the dynamics of the very early Universe. By this we mean epochs very much earlier than those we have considered so far. The earliest times we have discussed occurred at the time the protons and antiprotons annihilated at temperatures exceeding 10^{12} K. Although this is a very high temperature, it is still well within the range of energies which have been studied in detail in particle physics experiments. The processes we will discuss are assumed to have taken place at very much earlier times and at very much higher energies.

The most interesting concept for solving some of the fundamental problems is embodied in the idea of the *inflationary Universe*, which was introduced by Alan Guth in the early 1980s. Guth developed his brilliant ideas on the basis of the theory of elementary particles but the basic concepts can be understood, independent of that theory. The idea is that, in the very early Universe, the Universe expanded exponentially rapidly, driven by quantum mechanical forces, which have no counterpart in classical physics. Let us assume that such an expansion occurred so that the scale factor, R, increased exponentially with time, that is, as $R \propto e^{t/t_0}$. This means that the distance between any two points increases by the same factor in successive equal time intervals. Interestingly, such exponentially expanding models were found in some of the earliest applications of Einstein's equations to the Universe as a whole. The exponential expansion continues for a certain time and then the Universe switches over to the standard radiation-dominated phase of the early Universe. The epoch during which this exponential expansion of the Universe took place is called the *inflationary* era.

If this process were to occur, it would have some remarkable consequences. Consider a tiny region of the early Universe expanding under the influence of the exponential expansion. Particles within the region are initially very close together and so can communicate with each other. Before the inflationary expansion begins, there is time for the small region to become uniform and homogeneous. The region then expands at an exponentially increasing rate and so neighbouring parts of the region are driven to such distances that they can no longer communicate with each other by light signals – causally connected regions are swept beyond their local horizons by the inflationary expansion. At the end of the inflationary epoch, the Universe transforms into the standard radiation-dominated Universe and the inflated region continues to expand as $R \propto t^{\frac{1}{2}}$. To account for the observed isotropy of the Universe, the region which began as a tiny causally-connected region, is inflated by an enormous factor, roughly 10^{43} in size, to a dimension much greater than the present horizon scale of the Universe. This is an attractive way of accounting for the fact that regions which are no longer in causal contact can have the same properties, a solution to the horizon problem.

Another important effect of the exponential expansion is that it straightens out the geometry of the early Universe, however complicated it may have been to begin with. Again, suppose the little region of the early Universe has some complex geometry. The effect of the expansion is to increase the radius of curvature of the geometry by the same enormous factor as the exponential expansion. The little patch is inflated to dimensions vastly greater than the present size of the Universe and the geometry of the inflated region which becomes our Universe is driven towards flat Euclidean geometry. As explained in Section 4.3, in cosmology, the geometry of space–time and its energy density are very closely related and so, when the Universe transforms from the exponentially expanding inflationary state into the radiation-dominated phase, the geometry is Euclidean and consequently the Universe must have the critical density.

It is interesting that these arguments can be made quite independently of an understanding of the physics of the processes responsible for the inflationary expansion. Fortunately, physical forces which have the right properties have been discovered by the particle physicists. A key development in particle physics has been the introduction of what are called *Higgs fields* into the theory of weak interactions, that is, the type of force responsible for the decay of the neutron and the interactions of neutrinos. These Higgs fields were introduced into the theory of elementary particles in order to eliminate singularities in that theory and to endow the $W^{\pm}$ and Z^{0} bosons with mass. Precise measurements of the masses of these particles at CERN have confirmed that theory very precisely. Although the theory of electroweak forces is in excellent agreement with experiment, the particles associated with the Higgs fields, the *Higgs bosons* have not yet been detected. These are assumed to have greater masses than could be detected in the present generation of particle physics experiments. It is one of the great challenges for the next generation of particle accelerators to discover the Higgs bosons.

The Higgs fields have the property of being *scalar* fields and these are quite unlike the fields which describe electromagnetism and gravity. A general property of the scalar fields is that they have a negative energy equation of state, namely, $p = -\rho c^2$, where p and ρ are the pressure and mass density associated with the scalar fields respectively. The behaviour of matter with such an equation of state is quite unlike that of a classical gas, in which the greater the density of the gas at a given temperature, the greater the pressure. In a negative energy equation of state, the opposite happens in that the pressure is proportional to *minus* the density of the matter. In fact, it is not a pressure at all – it should be called a tension. It is easy to show that this is exactly the type of force field which would result in the exponential expansion of the very early Universe. Thus, the discovery of the Higgs bosons will be important, not only for particle physics, but for cosmology as well.

In the most popular version of the inflationary scenario for the very early Universe, fields similar to the Higgs fields are assumed to be associated with a phase transition in the very early Universe. According to grand unified theories of elementary particles, at high enough energies, the strong, electromagnetic and weak forces are unified and it is only at lower energies that they appear as distinct forces. The characteristic energies at which the grand unification is expected to take place are enormous, about $E \sim 10^{14}$ GeV, and occurred at a time only about 10^{-34} seconds after the Big Bang. In a typical realisation of the inflationary picture, the negative energy equation of state continues to drive the inflation from this time until the Universe was about 100 times older. During this period, there was an enormous release of energy associated with the phase transition and this heats up the Universe to a very high temperature indeed. At the end of the inflation era, the dynamics become those of the standard radiation-dominated Big Bang.

Let us put some figures into this argument. Over the period from 10^{-34} seconds to 10^{-32} seconds, all dimensions in the Universe, including the radius of curvature of its geometry, increased exponentially by a factor of about $e^{100} \approx 10^{43}$. The distance r over which particles could communicate at the beginning of the inflationary phase, what is called the *horizon scale*, was only $r \approx ct \approx 3 \times 10^{-26}$ m. Since particles within this horizon scale can communicate with each other, it is physically possible for density and other fluctuations to be smoothed out on all scales less than this dimension. This smoothed out region is then inflated to a dimension 3×10^{17} m by the end of the period of inflation. This dimension then scales as $t^{\frac{1}{2}}$, as in the standard radiation-dominated Universe so that the region would have expanded to a size of 3×10^{42} m by the present day – this dimension far exceeds the present dimension of the Universe which is about 10^{26} m. This history of the inflationary Universe is compared with the standard Friedman picture in Figure 5.7.

Another important contribution of particle physics concerns the solution of the baryon-asymmetry problem. Although there is symmetry between particles and antiparticles, in the sense that, for every particle, there is a corresponding antiparticle, particle physics experiments have shown that there is a slight asymmetry between them. Specifically, it is found that, in the decay of the neutral K^0 meson, there is a slight asymmetry between matter and antimatter. The K^0 meson should decay symmetrically into equal numbers of particles and antiparticles but, in fact, it doesn't – there is a slight preference for matter over antimatter. The importance of this result is that it indicates that there is an absolute distinction between matter and antimatter. According to the theory of this process, particles and antiparticles are

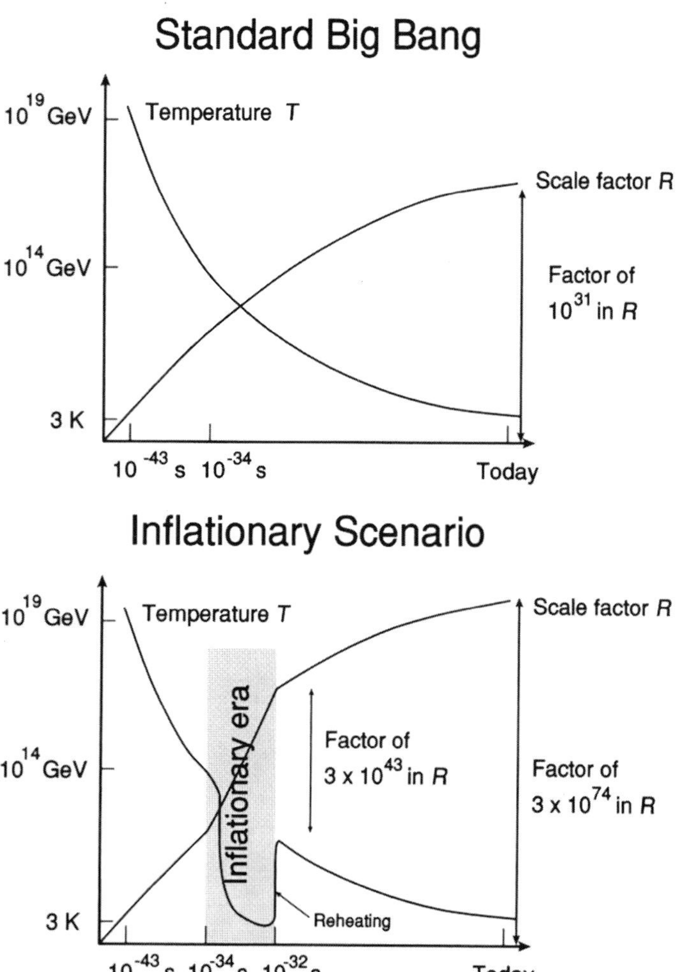

Figure 5.7. Comparison of the evolution of the scale factor and temperature in the standard Big Bang and inflationary cosmologies. The scale factor can be thought of as the distance between any two points which partake in the uniform expansion of the Universe.

indeed symmetrical at high enough energies but, at lower energies, the particles and antiparticles become distinct through the process known as spontaneous symmetry breaking. This idea can be built into the physics of the early Universe and theorists have found that roughly the observed asymmetry between matter and antimatter that we observe in the Universe can be explained.

The last problem concerns the origin of the fluctuations from which galaxies formed. There are several possible sources of primordial fluctuations in the scenario we have outlined. One possibility is that fluctuations are created during the phase change which was responsible for the inflationary expansion of the very early Universe. Whenever changes of state occur in nature, such as when water freezes or boils, there is the possibility of creating large fluctuations. Another possibility is that the fluctuations developed out of quantum fluctuations which must have been present before the inflationary era.

These ideas look promising but we should bear in mind that there is no evidence for the inflationary picture beyond the need to solve the four great problems. There is no question but that enormous progress has been made in understanding the types of physical process necessary to resolve the four great problems but it is not clear how we are to find independent evidence for the physical processes responsible for the inflationary expansion.

A representation of the evolution of the Universe from the Planck era to the present day is shown in Figure 5.8. The *Planck era* is that time in the very remote past when the energy densities were so great that a quantum theory of gravity is needed. On dimensional grounds, we know that this era must have occurred when the Universe was only about 10^{-44} seconds old. Despite enormous efforts on the part of theorists, there is no quantum theory of gravity and so we can only speculate about what the physics of these extraordinary eras must have been.

Figure 5.8 is drawn on a logarithmic scale and so we are able to encompass the whole of the Universe, from the Planck era at 10^{-44} seconds to the present age of the Universe which is about 3×10^{17} seconds or 10^{10} years old. Halfway up the diagram, from the time when the Universe was only about a millisecond old, to the present epoch, we can be reasonably confident that we have the correct picture for the Big Bang despite the four basic problems described above. At times earlier than about a millisecond, however, we very quickly run out of known physics. Indeed, in the model of the very early Universe we have been describing, it is assumed that we can extrapolate across the huge gap from 10^{-3} seconds to 10^{-44} seconds

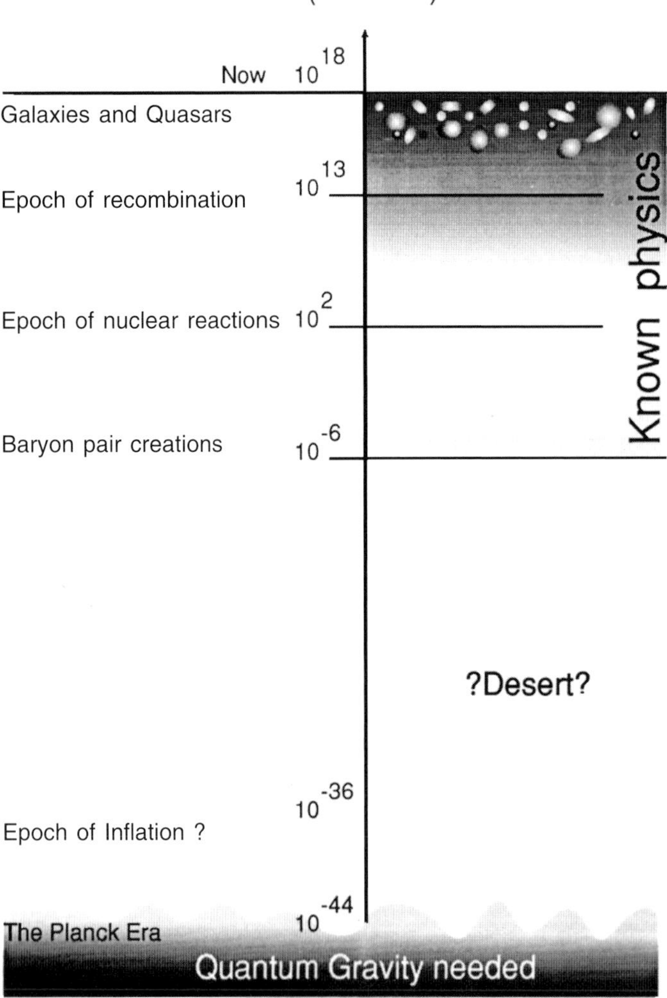

Figure 5.8. A schematic diagram illustrating the evolution of the Universe from the Planck era to the present time. The shaded area to the right of the diagram indicates the regions of known physics.

using our current understanding of laboratory physics. Maybe the current thinking will turn out to be correct but there must be some concern that some fundamentally new physics may emerge at higher and higher energies before we reach the Planck era at $t \sim 10^{-44}$ seconds.

The one thing which is certain is that at some stage a quantum theory of gravity is needed. Roger Penrose and Stephen Hawking have developed very powerful singularity theorems which show that, according to classical theories of gravity under very general conditions, there is inevitably a physical singularity at the origin of the Big Bang, that is, as $t \rightarrow 0$, the energy density of the Universe tends to infinity. One of the possible ways of eliminating this singularity is to find a proper quantum theory of gravity. This remains an unsolved problem and we can be certain that our understanding of the very earliest stages of our Universe will remain incomplete until it is solved. Thus, there is no question but that new physics is needed if we are to develop a convincing physical picture of the very early Universe.

5.9 Final things

As we have progressed through this chapter, we have continually run up against the limits of our current understanding of fundamental physics. Much of the speculation has gone far beyond what can possibly be validated by laboratory experiments. On the other hand, some of the concepts, such as the idea of inflation, may well have more general validity than the specific realisations which are currently being intensively studied. The situation is no different from any other of the physical sciences. At some stage, we run up against the limits of our physical theories and the challenge is to find ways of breaking through the barrier to new understandings which can be validated by experiment or observation. This means that there are many questions to which we simply cannot give convincing answers on the basis of accepted theory. We may guess what the answer is likely to be but that is very different from constraining concrete theories by real experiments and observations.

Despite the limitations of physics to provide answers, a number of unsettling questions crop up. If they can, the professional cosmologists avoid these issues, for good reasons, but let us try to grasp some of these nettles as best we can.

(1) *Are there other Universes outside the region which we call 'our Universe'?* If we interpret the Big Bang models literally, those with values of the density parameter, $\Omega \leq 1$ are infinite in extent. The models with $\Omega > 1$ have a finite size and have closed geometries. In particular, the popular critical model with $\Omega = 1$ is infinite in extent. We have, however, to emphasise the word 'model' in this discussion. The standard Big Bang picture seems to be an excellent description of the large scale features of our Universe and the evidence is remarkably convincing that the Universe went through a hot dense phase. It is, however, the simplest possible first approximation to the dynamics of the Universe. In principle, there might be quite separate Universes but we have no evidence for them. It is difficult to imagine how we would ever obtain any evidence for them in view of the photon and neutrino barriers which occur at times when the isotropic Big Bang seems to work very well indeed. Variants of the inflationary picture have been proposed in which multiple Universes might arise but these ideas again fall into the realm of speculation, unsupported by observational or experimental evidence.

(2) *What happens at the origin of the Big Bang?* According to the remarkable singularity theorems of Hawking and Penrose, there is a physical singularity in the standard world models at $t = 0$. According to the classical picture, the density and energy of the Universe diverge to infinity at that time. Classical physics is unable to cope with this singularity in space–time and we can regard time as beginning at that point.

(3) *What happened before $t = 0$?* According to classical physics, it is meaningless to ask the question because of the physical singularity at $t = 0$. Although we do not have a quantum theory of gravity, such a theory must alter the nature of the singularity theorems. At the Planck time, illustrated in Figure 5.8, space–time itself is quantised and theorists refer to the structure of the Universe at that time as being like a quantum 'space–time foam'. Maybe there is a way in which the Universe can evolve into such a state and then out of it into the Universe as we know it. At least, in principle, it is possible to conceive of the possibility that, once a satisfactory quantum theory of gravity is discovered, we may be able to evolve world models from the epochs $t \leq 0$ to our present Universe. But, how can we ever validate the theory?

(4) *What is the ultimate fate of the Universe?* According to the standard models, this entirely depends upon the density parameter Ω. In those models which expand forever, all nuclear sources of fuel will eventually be exhausted, leaving behind inert objects, such as black holes, brown drawfs and rocks. Matter, which can fall into black holes, will do so but most of it will probably escape that fate. Those models which have density parameter $\Omega > 1$ eventually collapse to a Big Crunch, as shown in Figure 4.9, which behaves like a time-reversed Big Bang. All the problems of treating times approaching $t = 0$ reappear as the Universe collapses towards the Big Crunch. Again, according to the classical picture, the Universe collapses to a physical singularity but what happens when the problem is treated quantum mechanically is not understood. If the Universe reverts to a 'quantum space–time foam', perhaps something re-emerges from this extraordinary state. We simply cannot answer this question at the moment.

In the past, when apparently insurmountable problems have been encountered, the resolution comes from some quite unexpected direction – some experiment, observation or theoretical discovery puts a completely new complexion upon the problem and the new insights gained indicate the way forward. Cosmologists are born optimists and my own view is that, despite the fact that many of these issues are at present beyond the power of physics to provide verifiable solutions, these will be forthcoming, but how this will come about cannot be foreseen. What is beyond question is that the way in which progress will be made is by the combination of advanced experiment, observation and theory. This is why the next generations of large telescopes for all wavebands and of large particle accelerators are so important – these are essential for the validation of theory and for uncovering new facts about the physical processes involved in the origin and evolution of our Universe.

I am often asked in popular lectures about the broader significance of these studies – are there implications for religion, theology and philosophy in the scientific pursuit of cosmology? In my view, physics provides us with a *description* of nature, which can be written down extraordinarily economically in mathematical terms. The miracle is that natural processes can be mathematized in this way. These tools are enormously powerful and they enable us to make quantitative predictions to new circumstances which can be validated by experiment. In the end, however, we are simply providing desciptions of the workings of nature and not *explanations*. Many of the great discoveries of physics have changed very profoundly some of our intuitive notions about the nature of our Universe. For example, the concept that space and time are not independent entities but are only part of four-dimensional space–time may not be intuitively obvious but that is how nature works. The fact that quantum mechanics tells us that there are fundamental uncertainties in the precision with which we can measure the position and velocity of any object is also non-intuitive but has been confirmed by myriads of experiments. These descriptions provide new insights into the behaviour of matter but they do not provide us with explanations of *why* matter behaves in these ways. In my view, this issue is outside the realm of physics.

The issue is beautifully encapsulated in a delightful interview I had with the chaplain of Trinity College, when I first arrived in Cambridge in 1963. I was invited to visit him for a chat and to find out if there was any way in which he could provide support. At that time, I was in the thick of the controversy about the number counts of extragalactic radio sources and described enthusiastically to him the evidence we were obtaining about the relative merits of Big Bang and steady state cosmologies. At the conclusion of our discussion, he made a remark which epitomises for me the distinction between what we do as professional cosmologists and its broader significance. He said very simply, 'Whatever the correct theory for the origin of our Universe, I never cease to wonder at the work of God's hands.' That seems to me to be a very healthy and proper attitude.

Further reading

There are many excellent introductory books on modern astronomy and cosmology. The following are among the books I refer to most often:

The Cambridge Atlas of Astronomy, 3rd Edition, eds. J. Audouze and G. Israel, 1994. Cambridge University Press. This book is profusely illustrated with many excellent images of all classes of objects in the Universe. The level is non-technical.

Images of the Cosmos by B. W. Jones, R. J. A. Lambourne and D. A. Rothery, 1994. Hodder and Stoughton, in association with the Open University. This is an excellent set of up-to-date images of all classes of objects in the Universe, prepared for the Open University courses in astronomy.

A View of the Universe by D. Malin, 1993. Sky Publishing Corporation and Cambridge University Press. A beautiful set of colour photographs of many of the most spectacular objects in the Universe by the master of astronomical colour photography.

The New Physics, ed. P. C. W. Davies, 1989. Cambridge University Press. The book contains a number of articles on general relativity, cosmology and astrophysics roughly at the level of *Scientific American*.

Fundamental Astronomy by H. Karttunen, P. Kroger, H. Oja, M. Pountanen and K. J. Donner, 1987. Springer-Verlag. This is a beautiful modern textbook on basic astronomy. It provides an excellent introduction at a slightly more technical level to many of the topics discussed in this book.

The Physical Universe: an Introduction to Astronomy by F. H. Shu, 1982. Mill Valley, CA. University Science Books. This is an excellent elementary textbook on astronomy at a basic undergraduate level.

The Stars: their Structure and Evolution, 2nd Edition, by R. J. Tayler, 1995. Cambridge University Press. This book is a semi-popular introduction to the physics of the stars. I have used the first edition as a primary reference in teaching elementary astronomy. It is a little gem of a book.

The Fullness of Space by C. G. Wynn-Williams, 1992. Cambridge University Press. An excellent non-technical introduction to the interstellar medium and the formation of stars.

Galaxies: Structure and Evolution by R. J. Tayler, 1993. Cambridge University Press. This is a semi-popular exposition of the properties of galaxies and the physical processes involved in their structure and evolution.

Black Holes and the Universe by I. D. Novikov, 1990. Cambridge University Press. This is a delightful popular introduction to the physics of black holes written by a leading expert in gravitation and cosmology.

High Energy Astrophysics, Volume 1 (1992) and Volume 2 (1994), by M. S. Longair. Cambridge University Press. The first two volumes of a three-volume series describing the tools needed for understanding modern high energy astrophysics. It is written at the final-year undergraduate level.

Principles of Cosmology and Gravitation by M. V. Berry, 1989. Adam Hilger. This is an elementary introduction to general relativity and cosmology at the undergraduate level.

Principles of Physical Cosmology by P. J. E. Peebles, 1993. Princeton University Press. A monumental survey of much of modern cosmology at the graduate level.

The Early Universe by E. W. Kolb and M. S. Turner, 1990. Addison–Wesley Publishing Co. A splendid introduction to the application of ideas of particle physics to the early Universe. It is at the graduate level.

Picture acknowledgements

1.1 From the edition of Copernicus' *De Revolutionibus Orbium Celestium* in the Crawford Collection at the Royal Observatory, Edinburgh
1.2 From J. Fraunhofer from *Denk. Münch. Akad. Wissen.* (1817)
1.3 From R. Learner *Astronomy through the Telescope* (1981). Evans Brothers Ltd, London
1.4, 1.11, 1.22, 2.24, 3.20, 4.13 Courtesy of NASA and the Space Telescope Science Institute
1.5, 1.9, 1.10, 2.11(a), 2.12, 2.14, 2.16(a), 3.9, 4.1 Courtesy of D. Malin and the Anglo-Australian Observatory
1.6 Courtesy of the Mount Palomar Observatory
1.7 Courtesy of the Lund Observatory
1.8, 1.29, 1.30, 4.6 Courtesy of NASA, the Goddard Space Flight Center and the COBE Science Team
1.12, 1.13 Courtesy of the Royal Greenwich Observatory
1.14 Courtesy of H. Arp from *Optical Jets in Galaxies* (1981), ESA Publications
1.15, 2.15, 2.16(b), 4.3(a), 4.15 Courtesy of the Royal Observatory, Edinburgh
1.16 Courtesy of P. J. E. Peebles from *Astron. J.* (1977)
1.17 Courtesy of J. Huchra and M. Geller of the Harvard–Smithsonian Center for Astrophysics
1.21 After R. Giacconi, H. Gursky and L. P. van Speybroeck, from *Ann. Rev. Astr. Astrophys.* (1968)
1.24 Courtesy of S. van den Bergh, from *Astrophys. J.* (1975)
1.25 Courtesy of F. Seward and the Harvard–Smithsonian Center for Astrophysics
1.26(a) Courtesy of NASA and the HEAO-1 Science team
1.26(b) Courtesy of J. Trümper and the Max Planck Institute for Extraterrestrial Physics, Garching
1.27 Courtesy of NASA, D. Kniffen and the Compton Gamma-Ray Observatory Science Team
1.28 Courtesy of NASA, the Jet Propulsion Laboratory and the Rutherford–Appleton Laboratory
1.31 Courtesy of G. Haslam and the Max Planck Institute for Radio Astronomy, Bonn
2.4 Courtesy of J. N. Bahcall, from *Neutrino Astrophysics* (1989), Cambridge University Press
2.5 Courtesy of K. S. Hirata *et al.*, from *Phys. Rev. Letters* (1990)
2.6(a) Courtesy of J. Kelly Beatty and A. Chaikin (eds), from *The New Solar System* (1990), Cambridge University Press
2.6(b), 2.25, 4.25 Courtesy of the US National Optical Astronomy Observatories
2.7 Courtesy of T. Toutain and C. Frölich from *Astron. Astrophys.* (1992)
2.8 Courtesy of D. O. Gough (1993)
2.9(a), (c) Courtesy of the European Space Agency, M. A. C. Perryman and the Hipparcos Science Team (1994)
2.10 Courtesy of D. Michalis and J. Binney from *Galactic Astronomy; Structure and Kinematics* (1981), W. H. Freeman and Co.
2.11(b) After J. E. Hesser *et al.*, from *Publ. Astron. Soc. Pacific* (1989)
2.13(a) Painting by Davis Melzer
2.19 Courtesy of the Royal Observatory, Edinburgh and K. Papworth, Cavendish Laboratory, Cambridge

2.22 After C. J. Lada, from *Ann. Rev. Astron. Astrophys.* (1985)
2.26 Courtesy of NASA, C. R. O'Dell and the Hubble Space Telescope Science Institute
2.27 Courtesy of F. H. Shu, F. C. Adams and S. Lizano, from *Ann. Rev. Astron. Astrophys.* (1987)
2.28 Courtesy of NASA, C. Burrows and the Hubble Space Telescope Science Institute
3.2, 3.3 Courtesy of R. Perley, copyright NRAO (1985) and *Astrophys. J.* (1984)
3.4(a) From *Structure and Evolution of Active Galactic Nuclei* (1986), D. Reidel and Company;
3.4(b) Courtesy of K. Pounds and the European Space Agency
3.5 Courtesy of A. Hewish, from *Q. J. R. Astron. Soc.* (1986)
3.7 After S. L. Shapiro and S. A. Teukolsky, from *Black Holes, White Dwarfs and Neutron Stars: The Physics of Compact Objects* (1983), Wiley-Interscience
3.8 Courtesy of A. R. Bell, from *Mon. Not. R. Astron. Soc.* (1977) and S. F. Gull
3.10 Courtesy of R. A. Chevalier from *Nature* (1992) and the Anglo-Australian Observatory
3.11 Courtesy of H. Tananbaum *et al.*, from *Astrophys. J.* (1972)
3.14, 3.15 Courtesy of J. Tayler, from *Proc. Roy. Soc.* (1992)
3.19 Courtesy of J. E. McClintock, from *Texas–ESO–CERN Symposium on Relativistic Astrophysics, Cosmology and Fundamental Physics* (1992), New York Academy of Sciences
3.21 After A. Wandel and R. F. Mushotzky, from *Astrophys. J.* (1986)
3.22(a) After T. J. Pearson *et al.*, from *Extragalactic Radio Sources* (1982), D. Reidel and Company
4.2 Courtesy of H. C. Arp and B. F. Madore from *A Catalogue of Southern Peculiar Galaxies and Associations* (1987), Cambridge University Press
4.3(b) Courtesy of A. and Yu. Toomre
4.4 After A. R. Sandage from *Observatory* (1968)
4.5 Courtesy P. J. E. Peebles *Principles of Physical Cosmology* (1993), Princeton University Press
4.10 From M. S. Longair, from *Alice and the Space Telescope* (1989), Johns Hopkins University Press, drawing by Stephen Kraft
4.12 Courtesy of H. Bondi from *Cosmology* (1960), Cambridge University Press
4.14 From S. J. Lilly and M. S. Longair in *Mon. Not. R. Astr. Soc.* (1984)
4.16 Courtesy of A. Stockton, from N. Henbest and M. Marten, *The New Astronomy* (1983), Cambridge University Press
4.17 Courtesy of Dr Simon Garrington and the MERLIN staff at Jodrell Bank, University of Manchester
4.18 Courtesy of NASA, J.-P. Kneib, R. Ellis and the Hubble Space Telescope Science Institute
4.19(b) Courtesy of C. Alcock *et al.*, from *Nature* (1994)
4.22 After C. Frenk, from *Phil. Trans. Roy. Soc. Lond.* (1986)
4.24 Courtesy of NASA, M. Dickinson and the Hubble Space Telescope Science Institute
5.2 Courtesy of R. V. Wagoner, from *Astrophys. J.* (1973)
5.3 Adapted from J. Audouze, *Astrophysical Cosmology* (1982), Scientiarium Scripta Varia, Vatican City
5.6 After F. Close, M. Marten and C. Sutton, from *The Particle Explosion* (1987), Oxford University Press
5.7 After E. Kolb and M. S. Turner, from *The Early Universe* (1990), Addison–Wesley Publishing Co.

Glossary

absolute zero The temperature at which all molecular motion ceases according to classical physics and so the lowest possible temperature. It is the zero point of the Kelvin temperature scale and corresponds to about −273 degrees Celsius.

absorption The process by which the intensity of radiation decreases as it passes through a material medium. The energy lost by the radiation is absorbed by the medium. Absorption associated with transitions in atoms or ions gives rise to absorption lines at specific wavelengths.

accretion The process by which matter accumulates onto a compact object under the influence of gravity. During star formation, gas is accreted onto the core of the star as its total mass is built up and the central temperature increases. Accretion of matter onto compact stars such as white dwarfs, neutron stars and black holes results in a large release of energy. Accretion of matter onto supermassive black holes in active galactic nuclei is the likely source of their enormous luminosities.

accretion disc If the material accreted onto a compact object possesses some rotation, or more precisely angular momentum, the material collapses to form a rotating accretion disc about the compact object and matter is then accreted onto it from the inner layers of the disc.

active galactic nucleus (AGN) The central region of an active galaxy in which exceptionally large amounts of energy are generated by non-stellar processes. This is a characteristic of various types of galaxy that have been classified in different ways according to their appearance and the nature of the radiations they emit. Quasars, Seyfert galaxies and radio galaxies are all manifestations of the AGN phenomenon. The source of energy is concentrated within the nucleus. One mechanism which can account for the huge amount of energy radiated involves the presence of a supermassive black hole, into which matter is accreted with the release of gravitational energy. Variability in the intensity of the radiation from the nucleus over very short time-scales shows that the energy source must be concentrated into a very small region of space.

Aitoff projection A geometrical projection by which it is possible to represent the whole of the sky on a single flat sheet of paper. In the most common version of this projection, the Galactic plane is oriented along the central axis of the diagram with the Galactic centre at its centre.

angular momentum The rotational momentum a body possesses by virtue of its rotation. An object with angular momentum continues to rotate at the same rate unless a couple or torque acts upon it – this principle is known as the conservation of angular momentum.

anthropic principle The idea that the presence of intelligent life on Earth influences the nature of the Universe we live in. According to the anthropic principle, there are only certain types of Universe in which intelligent life could possibly have developed. In other words, the Universe is as it is because we are here to observe it.

antimatter Matter composed of elementary particles which have masses and spins identical to the particles that make up ordinary matter, but with all other properties, such as electric charge, reversed. Although some antiparticles are observed in nature and others are produced in the laboratory, there is no evidence for the existence of antimatter in bulk as such. If matter and antimatter collide, they annihilate with the release of energy normally in the form of γ-rays.

Big Bang The standard model for the history of the Universe according to which it began in an infinitely compact state and has been expanding ever since that event, which took place about 10 to 20 billion years ago. The theory has been adopted by nearly all cosmologists as the framework for cosmological studies since it explains three of the most important observations in cosmology: the expansion of the Universe, the existence of the cosmic microwave background radiation and the origin of the light elements.

Big Crunch The ultimate fate of world models which eventually collapse to a compact dense phase. In many ways, the event is similar to a time-reversed Big Bang. In the simplest world models, this behaviour is found for models with density parameter Ω greater than 1 and deceleration parameter q_0 greater than $\frac{1}{2}$.

binary pulsar A pulsar which is a member of a binary star system. The most interesting cases are those systems which consist of two neutron stars, the first of these to be discovered being the system PSR1913 + 16, which consists of a radio pulsar and a neutron star. These objects provide important tests of theories of relativistic gravity.

binary star A pair of stars in orbit about their common centre of gravity, held together by their mutual gravitational attraction. About half of all stars are in fact binary or multiple systems, though normally the components cannot be distinguished individually. The presence of more than one star is inferred from the appearance of their combined spectrum. The two components in a binary system each move in elliptical orbits about their common centre of mass. Some binary stars are so close together that the pull of gravity distorts the individual stars from their normal spherical shape. They may exchange material and be surrounded by a common gaseous envelope. An accretion disc may develop about one of the stars due to mass transfer from the other star. The energy released results in the emission of ultraviolet and X-radiation. Novae are another consequence of mass exchange in binary stars.

bipolar outflow Gas streaming outwards in two opposite directions from a newly formed star. The outflow is assumed to originate from the inner regions of the star's accretion disc and moves out along the rotation axis. These stellar winds, moving at speeds up to 200 km s^{-1}, sweep up interstellar material, creating a double-lobed structure, extending outwards for distances of a light year or more.

black body A body that absorbs all the radiation incident upon it. The intensity of the radiation emitted by a black body and its spectrum depend only upon the temperature of the body. The spectrum of black-body radiation was first derived by Max Planck in 1900 and so it is often referred to as a Planck spectrum.

black hole An object which has undergone ultimate gravitational collapse. The force of gravity becomes so strong that, according to general relativity, matter and radiation which come too close to the black hole cannot be prevented by any known physical force from collapsing to a physical singularity at its centre. Matter and radiation cannot escape to the outside world from within the Schwarzschild radius of a non-rotating black hole. Black holes can possess only three properties, mass, electric charge and angular momentum.

brown dwarf A term describing objects with masses too small for nuclear reactions to be ignited in their cores. They are therefore very cool objects which can only be observed in the infrared waveband. Planets are examples of such objects. Stars with masses less than about 0.08 times the mass of the Sun are expected to be brown dwarfs. No brown dwarf has yet been unequivocally identified outside the Solar System, although a number of candidates have been suggested.

cataclysmic variable A close binary star system, normally involving the accretion of matter from a low mass main sequence star onto a white dwarf.

causality A basic principle of the theory of relativity according to which information cannot be transmitted between any two points at a speed greater than the speed of light. If this were possible, it can be shown that effect could precede cause in certain frames of reference. This would be a violation of causality.

cepheid variable A luminous, regular variable star with a characteristic light curve, that is, the variation of brightness with phase. The brightness increases rapidly to a maximum value and then decays more slowly. There is a very well defined relation between the luminosities of the cepheid variables and their periods so that, by measuring the period of a variable, its luminosity can be determined. This provides one of the best methods for measuring the distances of nearby galaxies.

cold dark matter A hypothetical form of dark matter which is normally assumed to consist of weakly interacting massive particles, which were present in equilibrium with all other forms of matter and radiation in the earliest phases of the Big Bang. Because they are assumed to be very massive, the particles would be very cold at the present time.

collimation The process of making a beam of light or particles parallel so that it is neither converging nor diverging.

continuum radiation A continuous spectrum of radiation as opposed to line emission or absorption. In the typical spectrum of a star, galaxy or active galactic nucleus, the emission and absorption lines are superimposed upon this continuous, or continuum, radiation spectrum. Black-body radiation is an example of continuum radiation, as is the synchrotron radiation of high energy electrons gyrating in a magnetic field.

cosmic microwave background radiation Diffuse electromagnetic radiation in the centimetre and millimetre waveband which pervades the whole of the Universe. Its discovery in 1965 was of immense importance for cosmology because it provided strong evidence in favour of the Big Bang model of the Universe. This background radiation is the relic of the hot early phases of our Universe. The spectrum of the background radiation is precisely that of a black body at a temperature of 2.725 degrees above absolute zero (2.725 K) and is most intense in the centimetre and millimetre regions of the spectrum. The radiation is quite remarkably uniform over the sky, the intensity being the same in all directions to a precision of about one part in 100 000. Our Galaxy is travelling through space at about 600 km s^{-1} relative to the background radiation.

cosmic rays High energy particles incident upon the top of the atmosphere. These particles consist of protons, electrons and nuclei and are present throughout the interstellar medium. They were probably accelerated in supernova explosions. Some of these particles are of quite enormous energy, up to 10^{20} eV.

cosmological constant A term added to the field equations of general relativity by Einstein in 1917 in order to create a static model Universe. The term corresponds physically to a repulsive force which counteracts the attractive force of gravity. Solutions of the field equations including the cosmological constant can result in model universes having ages greater than the inverse of Hubble's constant, unlike all those in which the cosmological constant is set equal to zero.

cosmological principle A basic principle of cosmology which states that our Galaxy is not located at any special position in the Universe. In other words, our Galaxy is at a typical location in the Universe and any other observer located on any other galaxy at the present time would observe the same large scale features of the Universe that we observe.

critical density In cosmology, the density of the critical world model, which just expands to infinity. The observed expansion is slowed down by the influence of gravity and can be reversed if the density is sufficiently high. The critical density is about $(1–2) \times 10^{-26}$ kg m^{-3}. This is about 100 times larger than the average density inferred to be present in visible matter, such as stars and galaxies. Many theorists believe that the Universe should have the critical density and so there is a dark matter problem, in the sense that there should be much more mass present in the Universe in forms we have not yet detected.

curved space Space in which the sum of the angles of a triangle do not add up to 180°. For example, if a triangle is drawn on the surface of a sphere, the angles of the triangle add up to more than 180°. Another example concerns drawing a triangle on a saddle, which is an example of hyperbolic space. The sum of the angles of the triangle in this case add up to less than 180°.

dark matter Matter which is present in the Universe but which emits very little radiation. Evidence for dark matter comes from observations of the velocities of galaxies within rich clusters of galaxies and from the velocities of rotation of giant spiral galaxies. A separate line of argument in favour of the presence of dark matter in the Universe comes from comparing the mass contained within galaxies with the critical cosmological density.

deceleration parameter The quantity q_0 which describes the rate at which the expansion of the Universe is slowing down. According to the simplest Friedman models, a value of q_0 greater than $\frac{1}{2}$ means that the expansion of the Universe will eventually be halted, followed by contraction and collapse. A value of q_0 less than $\frac{1}{2}$ indicates that the Universe will continue to expand forever.

degeneracy pressure The pressure associated with quantum mechanical forces which do not allow more than one particle of a particular type to occupy the same quantum mechanical state.

degenerate star A term which includes both white dwarfs and neutron stars, which are made up of degenerate matter. These stars are in an advanced state of evolution and have formed by the gravitational collapse of normal stars. Normal atoms cannot exist in their interiors under the conditions of very high pressure. In white dwarfs, the electrons and atomic nuclei form a dense, compressed mass. The degeneracy pressure of the electrons prevents the star collapsing under gravity. In neutron stars, the electrons and protons combine into a form of matter consisting of tightly packed neutrons and the neutron degeneracy pressure holds the star up against gravity. In both cases, there is an upper limit to the masses of these stars. For white dwarfs, the upper mass limit, known as the Chandrasekhar mass, is about 1.4 times the mass of the Sun and a similar value is found for the neutron stars. For greater masses, the star collapses to a black hole.

density contrast In cosmology, the density excess relative to the average background density in a particular region of space. It is believed that galaxies form out of these tiny density fluctuations which were formed in the very early Universe.

density parameter In cosmology, the density of the Universe relative to the critical cosmological density. It is written as the Greek letter Omega Ω.

Doppler effect The change of observed frequency of a source of radiation due to its motion. The difference between the emitted and received frequency is proportional to the speed of the source. If the source is moving towards the observer, the signal is received with a higher frequency; if the source if moving away from the observer, the observed frequency is less than the frequency emitted by the source in its own frame of reference. In the cosmological case, this increase in wavelength is referred to as the redshift of the radiation.

double quasar A pair of identical quasars very close together on the sky. These are the gravitationally lensed images of a single background quasar. The first double radio quasar 0957+561 was discovered in 1979.

dust grains Small particles of matter, typically about 100 nm in diameter, which co-exist with atoms and molecules of gas in interstellar space. The dust grains are thought to consist mainly of silicates and/or carbon in the form of graphite. Dark dust clouds are evident in the Milky Way where they obscure the light from stars and luminous gas clouds. Though tenuous, such clouds are very effective at absorbing visible light. Radiation at infrared and longer wavelengths can, however, pass through the dust clouds unimpeded. The presence of dust is also revealed by the emission of far infrared radiation, emitted when the grains are warmed by the absorption of visible and ultraviolet radiation. The temperature of dust is typically in the range 30–100 K.

Eddington limit An upper limit to the ratio of the luminosity to the mass of a stable star. Arthur Eddington showed that the limit is 40 000 when units of the Sun's mass and luminosity are used. If this limit is exceeded, the outer layers of the star are blown away by radiation pressure. The Eddington limit is also important in the study of X-ray binaries and active galactic nuclei and sets an upper limit to the luminosities which can be generated by accretion onto neutron stars and black holes.

Einstein ring The gravitationally lensed circular or elliptical image of a background object, when that object and the lensing galaxy are perfectly aligned.

electromagnetic radiation The radiation associated with oscillating electric and magnetic fields. Light is a form of electromagnetic radiation, as are radio, infrared, ultraviolet, X-ray and γ-radiations. Electromagnetic radiation travels at the speed of light through a vacuum. When charged particles are accelerated, they emit electromagnetic radiation.

epoch of recombination The epoch in the history of the Universe when the primordial plasma recombined to form a neutral gas. This epoch occurred at a redshift of about 1000, when the Universe was about 300 000 years old.

escape velocity The minimum velocity that will enable a small body to escape to infinity from the gravitational attraction of a more massive object.

Euclidean space Space that is uniform, homogeneous and isotropic and in which the angles of a triangle add up to 180°. Mathematically, the space has zero curvature and is the 'flat' space of everyday experience.

Friedman world models The standard solutions of Einstein's equations of general relativity for isotropic, homogeneous models of the Universe. In the simplest solutions, the cosmological constant is set equal to zero and the dynamics of the models can be thought of as a competition between the kinetic energy of expansion of the Universe, which makes the matter disperse, and the attractive force of gravity which tends to slow down the expansion. Friedman also found the solutions for the cases in which the cosmological constant is not zero.

general theory of relativity The relativistic theory of gravity discovered by Albert Einstein in 1915. The fundamental postulate of the general theory is the principle of equivalence, according to which it is impossible for an observer inside an enclosed elevator to distinguish between uniform acceleration of the elevator and a stationary elevator located in a uniform gravitational field. The theory describes gravitation as a geometric property of four-dimensional space–time. This geometry is in turn influenced by the mass and energy distribution. The general theory of relativity is important in many areas of modern astronomy and cosmology. The perihelion of Mercury's orbit precesses by 43 seconds of arc per century more than is predicted by Newton's theory of gravity but general relativity explains this difference exactly. The light from stars is deviated if the light rays pass very close to the Sun and this has been observed during solar eclipses. The motion of pulsars in binary systems provides some of the most important tests of general relativity and it has successfully passed all of them. The general theory of relativity is needed to understand the properties of black holes which are believed to be present in certain binary X-ray sources and in active galactic nuclei. General relativity is also essential for the construction of self-consistent models for the Universe as a whole. The standard Friedman models of the Universe are based upon the general theory of relativity.

giant branch The region occupied by stars towards the top right of the Hertzsprung–Russell diagram. The stars in this region of the diagram are cool but very luminous. Stars move onto the giant branch after they have completed their evolution on the main sequence. The lifetimes of stars on the giant branch are very much less than their main sequence lifetimes.

giant molecular cloud A cloud of interstellar matter containing molecules and interstellar dust. The most common molecule in giant molecular clouds is molecular hydrogen but it is very difficult to observe and the most common tracer of the distribution of molecules in the clouds is carbon monoxide, CO, through its millimetre line emission. The clouds contain many other molecular species. Giant molecular clouds are among the most massive entities within our Galaxy, with masses up to about 10 million times that of the Sun, and are typically 150 to 250 light years across. The Orion Giant Molecular Cloud is a notable example of such clouds.

gravitational lens Gravity bends the paths of light rays and so, when light passes by a massive object, the light rays are deflected by its gravitational influence. The result is that the light from the background star is deflected and focused by the intervening object. This gravitational lensing effect distorts the image of the background object and these distortions can be used to determine the mass distribution in the lensing object.

gravitational radiation Just as electromagnetic radiation is associated with the acceleration of electric changes, so when masses are accelerated relative to other masses there is, in general, the gravitational equivalent of electromagnetic radiation which is called gravitational radiation. Gravitational radiation is, however, very much weaker than electromagnetic radiation because of the difference in the strengths of the electromagnetic and gravitational forces. There has not yet been a direct detection of gravitational radiation by gravitational wave detectors but convincing evidence has been found for gravitational wave energy loss in the binary pulsar PSR1913+16.

gravitational redshift The redshift of electromagnetic radiation as it propagates away from a massive body. When radiation escapes from a deep gravitational potential well, such as those associated with a neutron star or a black hole, it has to do work against gravity and this leads to a decrease in the frequency of the electromagnetic waves.

helioseismology or solar seismology The study of the internal properties of the Sun by measuring its different resonant modes of oscillation. These modes of oscillation are constantly being excited by convective motions within the Sun.

Hertzsprung–Russell diagram or H–R diagram A diagram in which the luminosities of stars are plotted against their colours or spectral types. In the conventional way in which this diagram is plotted, luminosity increases logarithmically up the vertical axis and temperature increases from right to left along the horizontal axis. Stars do not occupy all regions of the H–R diagram but form various sequences, the most important being the main sequence, the giant branch and the horizontal branch.

hierarchical clustering The process by which objects are built up by the coalescence of smaller objects. The process is sequential so that a hierarchy of clustering is created by the coalescence of larger and larger structures. This is the process by which galaxies and larger scale structures are believed to be built up in the cold dark matter picture of galaxy formation.

Higgs particles Particles predicted by particle theorists to account for the masses of certain elementary particles. The success of the theory of the electroweak interaction in accounting for the masses of the $W^{\pm}$ and Z^0 particles is convincing evidence that these particles exist but they have not yet been observed in the most powerful accelerators. These particles have the important property of being described by scalar fields which are known as Higgs fields and fields of this nature are needed to drive the inflationary phases of the early Universe.

horizontal branch stars Luminous stars found in the H–R diagrams of globular clusters. It is believed that they originate from red giant stars which lose mass from their outer layers and so move to the left across the H–R diagram. As these stars evolve, they move back towards the red giant branch at more or less constant luminosity.

hot dark matter A possible variant of dark matter theories for the origin of structure in the Universe. In its most popular guise, it is assumed that the neutrinos have a finite rest mass of about 10 eV. There are sufficient relic neutrinos produced in the Big Bang that, if they had this mass, the total mass density in neutrinos would be sufficient to close the Universe. Since the neutrinos are assumed to have very small rest masses, the neutrinos remained hot until late in the Universe.

Hubble's constant The constant of proportionality H_0 in Hubble's law, $v = H_0 r$. It describes the rate at which the Universe expands at the present time. The value is not easy to determine, on account of uncertainties in the extragalactic distance scale, but it probably lies between 50 and 100 km s^{-1} Mpc^{-1}. In the standard models of the Universe, its value changes with time, rather than being a fundamental constant of our Universe.

Hubble's law The relation which describes the observation that the recession velocities of distant galaxies v are directly proportional to their distances r from our Galaxy, $v \propto r$. The law is a consequence of the uniform expansion of the Universe as a whole.

hyperbolic space Space that is uniform and homogeneous but in which the spatial sections are hyperbolic rather than flat. The simplest example of a hyperbolic space is the surface of a saddle. If a triangle is drawn on a saddle, the sum of the angles of the triangle is less than 180°. Mathematically, the space has negative curvature.

inflationary Universe A model for the early evolution of Big Bang models involving the exponential expansion of the Universe. This hypothetical phase in the evolution of the early Universe has been introduced to account for the observed isotropy of the Universe on a large scale and the fact that its density is within a factor of ten of the critical cosmological density. In the most popular version of the theory, the exponential expansion is associated with a phase change which occurred about 10^{-34} seconds after the beginning of the Big Bang. According to grand unified theories of elementary particles, the strong force decoupled from the electroweak force at this time. This event released enormous energy, stored until then in the vacuum of space–time. This scenario can account for the present vast extent of the Universe and its uniformity.

intergalactic medium The medium in the space between the galaxies. In rich clusters of galaxies, it is known that there is a very large amount of very hot gas in the space between the galaxies. It is assumed that there must be some matter in the space between the galaxies in general but it has proved very difficult to detect. Recently, ionised gas in the general intergalactic medium at a temperature of about 30 000 K has been reported at large redshifts.

interstellar chemistry Chemistry taking place in the rarified conditions of interstellar space. In giant molecular clouds, the number densities of particles are very much lower than those found in the best laboratory vacua. Since the interiors of the clouds are shielded from the dissociating effect of the ionising radiation of stars, molecules can be formed and a wide range of different chemical species are found in these regions.

interstellar medium The medium present in the space between the stars in galaxies. It consists of gas at a very wide range of densities and temperatures, from high temperature gas at about 10 000 000 K expelled in supernova explosions to cold gas in giant molecular clouds which can be as cold as about 10 K. The medium is threaded by a magnetic field and high energy particles, including protons, electrons and nuclei, permeate the interstellar medium.

ionisation The process in which electrons are removed from atoms either by collisions with other particles or by energetic photons. Since the electrons are negatively charged, the atoms become electrically charged and are known as positive ions, while the electrons are liberated. The resulting ionised gas consisting of positively charged ions and negatively charged electrons is known as a plasma.

isotropy The property of having no preferred direction. Liquid water is isotropic whereas a snowflake, which has sixfold symmetry, is not. The Universe on the largest scales is observed to be isotropic in all directions.

Jeans instability The instability, discovered by James Jeans in 1902, associated with the collapse of gas clouds under the influence of gravity. If the pressure within the cloud is not sufficient to counteract the inward pull of gravity, the cloud collapses. This process is important in the formation of stars.

Kerr black hole A rotating black hole. Black holes can rotate up to a certain maximum rotational speed. Maximally rotating Kerr black holes are important in high energy astrophysics because the maximum amount of energy can be liberated when matter is accreted onto them.

last scattering surface The surface of an object at which radiation was last scattered before it is observed on the Earth. In the case of the Sun, this surface corresponds to the photosphere of the Sun. In the case of the Universe, the surface lies at a redshift of almost exactly 1000. Prior to the epoch corresponding to this redshift, there were sufficient free electrons present in the intergalactic medium to scatter the cosmic background radiation many times. The photons of the cosmic microwave background radiation which we observe today were last scattered at a redshift of about 1000.

last stable orbit About all types of black hole, there is a last stable circular orbit. Circular orbits within this radius are unstable and the particles inevitably fall within the 'black' surface of the black hole. In the case of Schwarzschild (non-rotating) black holes, the last stable orbit lies at three times the Schwarzschild radius. For maximally rotating black holes, the last stable orbit, for particles orbiting in the same sense as the rotation of the black hole, is only half the Schwarzschild radius.

M numbers, Messier catalogue A catalogue of about 100 of the brightest galaxies, star clusters and nebulae, compiled by the French astronomer Charles Messier. His initial list, published in 1774, contained 45 objects but it was supplemented later with additional discoveries and contributions from Messier's colleague, Pierre Méchain. Objects in the catalogue, which is still widely used, are identified by the prefix 'M' and their catalogue number.

magnetic flux freezing A phenomenon which occurs when there is a magnetic field present in a plasma. The magnetic field becomes 'frozen' into the plasma, in the sense that, if the fluid moves, the magnetic field lines move as well, as if they were frozen into the plasma.

magnitude A measurement of the brightness of a star or other celestial object. On the magnitude scale, the lowest numbers refer to the objects of greatest brightness. The magnitude system was initially a qualitative attempt to classify the apparent brightness of stars. In about 120 BC, the Greek astronomer Hipparchus ranked stars on a magnitude scale from 'first' for the brightest stars to 'sixth' for those just detectable in a dark sky by the unaided eye. This qualitative description was standardised in the mid 19th century. By this time it was understood that each magnitude step corresponded roughly to a similar brightness ratio. In other words, the magnitude scale is a logarithmic scale of brightness. If S is the flux density, or brightness, of the object, its magnitude m is defined by an expression of the form $m = \text{constant} - 2.5 \log_{10} S$. The brightnesses of stars or galaxies as observed from the Earth, that is, their apparent magnitudes, depend on both their intrinsic luminosity and their distance. Absolute magnitude is a measure of intrinsic luminosity of the object on the magnitude scale. It is defined as the apparent magnitude the object would have if it were placed at a standard distance of 10 pc.

main sequence If a sample of stars is plotted on a Hertzsprung–Russell diagram, that is, a plot of stellar luminosity against stellar temperature, it is found that most of the stars in the Universe lie along a broad track stretching from high luminosity, high temperature stars at the top left, to low luminosity, low temperature stars at the bottom right of the diagram. This broad band is known as the main sequence. In main sequence stars, the fusion of hydrogen into helium in the stellar core is the source of energy. Cool stars with high luminosities, such as red giant stars, and hot stars with low luminosities such as white dwarfs are not main sequence stars.

main sequence termination point As a cluster of stars evolves, the most massive stars complete their evolution on the main sequence first and so, as the cluster ages, stars in the upper part of the main sequence are gradually depleted. The point on the main sequence above which there are no more stars is known as the main sequence termination point and the location of this point on the H–R diagram of a cluster provides a measure of its age.

massive compact halo object or MACHO A possible form of dark matter which could be present in the halo of our Galaxy, where it is known that some dark matter must be present. The MACHOs might be planets, brown dwarfs, neutron stars or black holes. These are very faint objects but they can be detected by the gravitational lens effect when they pass across the line of sight between a background star and the observer. The lensing by the MACHO results in a characteristic brightening of the background star.

matter-dominated Universe The late stages of evolution of the Universe when its dynamics are determined by the total mass density in matter rather than by that of the background radiation. Our Universe has been matter-dominated from about a redshift of 10 000 to the present day.

neutrino astronomy The attempt to detect neutrinos from cosmic sources, especially the Sun. Neutrinos are elementary particles with no electric charge, which interact only very weakly with other matter. They travel at the speed of light and are produced in vast quantities by the nuclear reactions that take place in the centres of stars and in supernova explosions. Neutrinos created in the nuclear reactions which take place in the centre of the Sun have been detected. In addition, a burst of neutrinos was detected when the supernova SN1987A exploded.

neutron star A star that has collapsed to such an extent under gravity that it consists almost entirely of neutrons. Neutron stars have radii of only about 10 km and have typical densities of about 10^{17} kg m^{-3}. They can be considered to be giant nuclei consisting of about 10^{60} neutrons.

NGC numbers, New General Catalogue A catalogue of non-stellar objects compiled by J. L. E. Dreyer of Armagh Observatory and published in 1888. It listed 7840 objects. A further 1529 were listed in a supplement that appeared seven years later, called the *Index Catalogue* (IC). *The Second Index Catalogue* of 1908 extended the supplementary list to 5386 objects. The NGC and IC numbers are widely used as a means of identifying nebulae and galaxies.

occultation The passage of one astronomical object directly in front of another so as to obscure it from view as seen by an observer on Earth.

Olbers' paradox The paradox noted by Olbers and earlier scientists which results from the observation that the sky is dark at night. If the Universe were infinite, static and uniformly filled with stars, the sky should be as bright as the surface of the stars. Since this is not the case, at least one of the above assumptions must be wrong. This is sometimes called 'the oldest fact in cosmology'.

parallax The apparent movement of a nearby star against the background of distant stars due to the Earth's motion about the Sun. This is the best method of measuring the distances of nearby stars.

Planck era The epoch in the extremely early Universe when it is necessary to take account of the effects of quantising gravity. Although there does not exist a quantum theory of gravity, on general grounds, it is expected that these effects become important at times about 10^{-44} seconds after the Big Bang.

planetary nebula An expanding shell of gas surrounding a star at a late stage of stellar evolution. The name derives from the description given by William Herschel, who thought their circular shapes were reminiscent of the discs of the planets as seen through a small telescope. There is no connection between planets and planetary nebulae. Planetary nebulae are formed as a result of mass loss at the end of evolution of red giant stars up the giant branch. A shell of gas is expelled and the cores of the dying stars ultimately become white dwarfs. Planetary nebulae take a variety of forms – ring-shaped, circular, dumbbell-like or irregular. Notable examples include the Ring Nebula, the Helix Nebula and the Dumbbell Nebula.

plasma A state of matter in which the atoms of a gas are ionised, so that the plasma consists of a gas of electrons, protons and ions of heavier elements. This state of matter has quite different properties from that of a neutral gas; in particular, it has very high electrical conductivity. Normally, cosmic plasmas have temperatures greater than about 10 000 K.

polarisation In electromagnetic waves, the electric field direction is always perpendicular to the direction of propagation of the waves. In unpolarised radiation, there is no preferred direction of the electric field vectors in the plane perpendicular to the direction of travel of the waves. In linear polarised radiation, all the vectors representing the electric fields of the waves are lined up at a particular angle in the plane perpendicular to its direction of travel. In circularly polarised radiation, the direction of polarisation changes continuously in such a way that the electric field vector rotates at the frequency of the wave.

primordial nucleosynthesis The synthesis of the elements by nuclear reactions in the early stages of the Big Bang. As the primordial plasma cools down from a very high temperature in the very early Universe, reactions between protons and neutrons become possible and, within the first few minutes of the evolution of the Universe, elements such as helium-4, helium-3, deuterium and lithium are synthesised. There is no other plausible origin for the observed abundances of these elements.

protostar A star in the earliest stages of formation as it condenses out of an interstellar cloud but before the onset of nuclear reactions in its core.

pulsar or pulsating radio source A radio source, characterized by regular bursts of extremely high brightness radio emission with a repetition frequency of between about 0.001 to 4 seconds. The parent bodies of pulsars are rotating, magnetised neutron stars, with masses similar to that of the Sun but with radii of only about 10 km. The radio pulses are associated with narrow beams of radio emission emitted along the magnetic axis of the neutron star and these are observed as pulses by the observer on Earth because the neutron star is rotating very rapidly.

quasar or quasi-stellar radio source A small extragalactic object that is exceedingly luminous for its angular size and has a large redshift. Quasars are the most luminous type of active galactic nuclei (AGN). In a small number, the faint nebulous light of a surrounding galaxy has been detected. Many thousands of quasars are catalogued. In general, quasars have a spectrum that shows emission lines, high redshifts (typically from 0.5 to 4, although higher and lower values than these are recorded) and they are so compact that they appear to be stars on photographs. Although the first quasars discovered in the 1960s were all radio sources, the majority of those now known are not strong radio sources.

radiation-dominated Universe The early phases of the Universe, when the dynamics were dominated by the mass density of radiation, rather than by that of the matter. At a redshift of about 100 000, the mass density of the cosmic microwave background radiation was equal to that of the critical model of the Universe. At earlier times, the mass density in the radiation exceeded that in the matter and so the dynamics were determined by the radiation rather than by the matter.

radio galaxy A galaxy which is an intense source of radio emission. About one galaxy in a million is a powerful radio galaxy. The radio emission is the synchrotron radiation of ultrarelativistic electrons travelling at speeds very close to the velocity of light in the magnetic fields in the radio source. The radio galaxy Cygnus A is often regarded as the prototype, in which two huge lobes of radio emission are disposed symmetrically on each side of a giant elliptical galaxy. Radio galaxies are closely related to quasars, many of which have similar radio properties.

radioactive decay The spontaneous decay of unstable elements created in supernova explosions and in other sites of nucleosynthesis.

recessional velocity The velocity with which a galaxy is observed to move away from our Galaxy due to the uniform expansion of the Universe.

red giant A star which is a member of the red giant branch. Once a star has moved off the main sequence, its core collapses and its outer envelope expands by an enormous factor. What is observed is a very large cool but luminous star which is known as a red giant.

redshift The shift of features in the spectrum of an astronomical object to longer wavelengths. If λ_{em} is the emitted wavelength of the radiation and λ_{obs} is the observed wavelength, the redshift z is defined to be the increase in wavelength, $\lambda_{obs} - \lambda_{em}$, divided by the emitted wavelength λ_{em}, that is,

$$z = \frac{\lambda_{obs} - \lambda_{em}}{\lambda_{em}}$$

In the case of galaxies, the redshift is due to the recessional velocity of the galaxy. The other astronomical cause of redshift is the gravitational redshift (see above).

relativistic beaming The beaming of the emission of sources of radiation which move at speeds approaching the speed of light due to their relativistic motion. When a source of radiation moves at such high speeds, even if the radiation is emitted isotropically in the frame of reference of the source, it is observed to be beamed very strongly in the direction of motion of the source. If the observer is located at a small angle to the direction of the beam, the intensity of the radiation is very greatly enhanced and the source can appear to move at a speed greater than that of light, known as superluminal velocity.

scale factor The quantity R which measures how the average distance between galaxies changes as the Universe expands. In the notation used in this book, the scale factor is taken to be 1 at the present day. When the galaxies were on average all closer together by a factor of 2, the scale factor was one half, $R = 0.5$; when they were on average four times closer together than at present, the scale factor was one quarter, $R = 0.25$, and so on.

Schwarzschild radius The critical radius at which the space–time surrounding a point mass becomes so curved that it wraps round to enclose the point mass. An object that has collapsed inside its Schwarzschild radius is a black hole, from which neither matter nor radiation can escape to the outside world. The Schwarzschild radius of the Sun is 3 km and of the Earth, 1 cm. The Schwarzschild radius is proportional to the mass M of the black hole and is given by the formula

$$R_S = 3\left(\frac{M}{M_\odot}\right) \text{km}$$

where $M_\odot$ is the mass of the Sun.

scintillation As applied to radio astronomy, the term scintillation means fluctuations in the intensity of the signal from the radio source due to irregularities in the medium through which the radio waves travel. The phenomenon is similar to the twinkling of stars. In both cases, the variations in the signal intensity have nothing to do with variations intrinsic to the source itself.

seeing The blurring of the optical image of a star due to irregularities along the paths which the light signals travel through the atmosphere to the telescope. Because of the problems of seeing, large optical telescopes never achieve the full resolving power which should be theoretically possible. For example, a 4-metre optical telescope should be able to resolve objects only 0.03 arcsec apart but the effects of seeing in the Earth's atmosphere degrade the resolution to only about 1 arcsec.

Seyfert galaxy A type of active galaxy with a brilliant point-like nucleus, first described by Carl Seyfert in 1943. Often the intensity of the nucleus is variable, indicating that it must be very compact. Their optical spectra contain intense emission lines. In Seyfert I galaxies, the lines are very broad whereas in Seyfert II galaxies the emission lines are much narrower.

singularity As applied to cosmology and black holes, a region of space where the the density becomes infinite. In these cases, the singularity can be visualised as a warped region of space–time where one or more of the quantities describing the geometry become infinite so that ordinary laws of physics cease to apply. According to the classical picture, the Big Bang emerged from such a singularity.

solar wind The outflow of hot gas from the corona of the Sun. At the distance of the Earth, the speed of the solar wind is about 350 km s^{-1}.

special theory of relativity The relativistic theory of space–time discovered by Einstein in 1905. In it, Einstein assumed that the laws of physics are the same in all frames of reference that are in uniform motion relative to each other, and that the speed of light is a fundamental constant of nature which takes the same value in all frames of reference. The equations of special relativity describe exactly how to relate the properties of objects moving at high speeds to what is observed by a stationary observer. The theory also leads to the equivalence of mass and energy through the famous equation $E = mc^2$ which states that any form of energy has a certain mass and equally a given amount of mass corresponds to a certain amount of energy.

starburst galaxies Galaxies in which a large burst of star formation is observed. The characteristic signature of these galaxies is intense emission in the far infrared waveband.

Stefan–Boltzmann law The law of physics which describes the total amount of radiation emitted by a black body maintained at a fixed temperature. The total emission is proportional to the fourth power of the temperature of the black body, $I \propto T^4$.

superluminal motion Apparent motion of an object at a velocity which exceeds that of light. The angular separation of the components of some compact double radio sources increase at a rate that is equivalent to as much as ten times the speed of light. It is probable that the effect is a geometrical illusion caused by the fact that the ejected component is travelling almost directly towards us along the line of sight at a velocity which is very close to the speed of light. The phenomenon has been observed in the quasar 3C 273 and many other compact quasars.

supernova A catastrophic stellar explosion in which so much energy is released that the supernova alone can outshine an entire galaxy of billions of stars. In addition to the light energy produced, ten times as much energy goes into the kinetic energy of the material blown out by the explosion and a hundred times as much is carried off by neutrinos. One way in which a supernova explosion can occur is when a massive evolved star has exhausted its nuclear fuel. Under these circumstances, the core becomes unstable and collapses in less than a second. The implosion may not, however, continue indefinitely. When the density of matter reaches nuclear densities, there is strong resistance to further pressure, the core bounces and an outward shock wave is generated. The outer layers of the star are blown outwards at thousands of kilometres per second, leaving behind a neutron star. Another way in which a supernova explosion might occur is if matter is continuously accreted onto a white dwarf in a binary star system. When the mass of the accreting star exceeds the Chandrasekhar limit, it collapses releasing the binding energy of the neutron star which forms.

supernova remnant The expanding shell of material created by the ejection of the outer layers of a star that explodes as a supernova. Well-known examples of supernova remnants are the Crab Nebula, Cassiopeia A, Kepler's supernova, Tycho's supernova and the Cygnus Loop.

synchrotron radiation Electromagnetic radiation emitted by ultrarelativistic electrons, that is, electrons travelling almost at the speed of light, through a magnetic field. The name arises because it was first observed in the synchrotron accelerators used by nuclear physicists. It is the major source of radio emission from supernova remnants and radio galaxies. Much of the light and the X-ray emission from the Crab Nebula is also produced by the synchrotron process by the very high energy electrons accelerated by the central pulsar. The spectrum of synchrotron radiation is very different from that of the thermal radiation emitted by hot gas. Sources of synchrotron radiation are thus easy to identify. The polarisation of the emission provides a means of estimating the strength of the magnetic field in the source region.

ultrarelativistic particles Very high energy particles, having kinetic energies which are very much greater than their rest mass energies.

very long baseline interferometry or VLBI The radio astronomical technique of combining the signals from radio telescopes separated by very great distances to produce very high definition images of radio sources. Typically, the angular resolution of such observations is about 0.001 arcsec.

voids Large holes in the distribution of galaxies on scales very much greater than those of clusters of galaxies. The largest voids can have dimensions of about 50 Mpc or greater.

white dwarf A star in an advanced state of stellar evolution, in which the pressure support is provided by the degeneracy pressure of the electrons. A white dwarf is created when a star like the Sun finally exhausts its sources of fuel for thermonuclear fusion. The star collapses under its own gravity, compressing the matter to a degenerate state in which electrons and nuclei are packed together.

Wien's displacement law The law which describes how the spectrum of black-body radiation changes with temperature. Specifically, the frequency at which the maximum intensity of the radiation is emitted is proportional to the temperature of the black body.

WIMPs Acronym for electrically neutral weakly interacting massive particles. These types of particles have been predicted by theories of elementary particles but they have not yet been observed in particle accelerators. They are important in cosmology because they are among the more promising candidates for the dark matter in galaxies and clusters and might well be present in sufficient quantities to close the Universe.

X-ray binary sources Compact X-ray sources which are members of binary star systems. The compact object may be a neutron star or black hole and the origin of the X-ray emission is accretion from the primary star onto the compact secondary.

Astronomical Units

The following conventions are used for the units throughout this book. SI (Système International) units are used which are based upon the metre (m) as the unit of length, the kilogram (kg) as the unit of mass and the second (s) as the unit of time. Some of the derived units used in the text are as follows.

Units of distance

In astronomy, the conventional unit is the parallax-second, which is abbreviated to parsec (pc). It is the distance at which the mean radius of the Earth's orbit about the Sun subtends an angle of 1 arcsec. It corresponds to a distance of 3.086×10^{16} m, that is, about 3×10^{16} m. A similar unit is the light year which corresponds to 9.4605×10^{15} m, which is about one-third of a parsec. Another useful astronomical distance measure is the astronomical unit which is defined to be the mean distance between the Sun and the Earth, that is, 1 astronomical unit (AU) = 1.496×10^{11} m.

Unit of energy

The unit of energy in the SI system is the joule but it is permissible to use the unit of the electronvolt to describe the energy of high energy photons, or the particles of electromagnetic radiation. For example, it is conventional to describe the energy of X-ray photons in terms of electronvolts (eV) or kiloelectronvolts (keV), which means 1000 electronvolts. Correspondingly, energetic γ-rays are often referred to in terms of megaelectronvolts (MeV) meaning 1 000 000 eV.

Unit of frequency

The unit of frequency is one cycle per second which is known as 1 hertz (Hz). Radio frequencies are described in megahertz (MHz), meaning 1 000 000 hertz or 10^6 hertz, and gigahertz (GHz) meaning 1 000 000 000 hertz or 10^9 hertz.

Unit of wavelength

In optical and infrared astronomy, it is conventional to use nanometres (nm), meaning one billionth of a metre, that is, 1/1 000 000 000 m or 10^{-9} m and microns (μm), meaning one millionth of a metre, that is, 1/1 000 000 m or 10^{-6} m.

Masses and Luminosities

It is often useful to relate the masses and luminosities of astronomical quantities to those of the Sun. The symbol $\odot$ is used to refer to the Sun. Then,

- Mass of Sun = $M_\odot = 1.989 \times 10^{30}$ kg $\approx 2 \times 10^{30}$ kg
- Luminosity of Sun = $L_\odot = 3.90 \times 10^{26}$ W
- Radius of Sun = $R_\odot = 6.9598 \times 10^8$ m

Index

The principal references to items in the index are indicated by bold page numbers. References to items in the Glossary are indicated in italic type.